THE
HANDBOOK
OF
INSECT
COLLECTING

THE
HANDBOOK
OF
INSECT
COLLECTING

THEIR COLLECTION,
PREPARATION, PRESERVATION
AND STORAGE

Courtenay Smithers

ANGUS
& ROBERTSON
PUBLISHERS

ANGUS & ROBERTSON PUBLISHERS

Unit 4, Eden Park, 31 Waterloo Road,
North Ryde, NSW, Australia 2113, and
16 Golden Square, London W1R 4BN,
United Kingdom

First published 1981 by A.H. & A.W. Reed Pty Ltd
First published in Australia
by Angus & Robertson Publishers in 1988
First published in the United Kingdom
by Angus & Robertson (UK) in 1988

National Library of Australia
Cataloguing-in-publication data.

Smithers, C.N. (Courtenay Neville), 1925-
  Handbook of insect collecting.

  Bibliography.
  Includes index.
  ISBN 0 207 15911 4.

  1. Insects — collection and preservation.
  1. Title.

579

Typeset by B & D Modgraphic Pty Ltd, Adelaide
Printed in Singapore

PREFACE

Insects have always fascinated people. There are more laymen interested in insects than any other group of animals with the possible exception of birds. Many insects are beautiful; some are bizarre; they are abundant; they are easily preserved and take up little space; collecting them takes you out of doors, adds point to your ramblings in pleasant places; their way of life is so different from our own that you enter a new world when you look at them with an interested eye and an open mind. Mankind's natural curiosity and appetite for the chase are both satisfied by the collection and study of insects.

Although insects are our greatest competitors for the resources of this world, this book deals with the study of insects for its own sake and not with a view to destroying them. Some of the rarer species are seriously threatened, if not driven to extinction, by the use of pesticides and by the destruction of their natural habitats as man gradually erodes the countryside. Indiscriminate collecting and 'trophy hunting' are of no scientific value and cause unnecessary disturbance of the Earth's ecological balance.

Most professional entomologists started their careers by making an insect collection. The world's greatest naturalists have all been collectors and many of the finest and scientifically important collections were started in a very modest way out of a casual interest. No matter how small they may be, most properly made collections will have something in them which is of interest now or will be in years to come.

Much of the pleasure of collecting comes from discussing your specimens with other collectors. There are several societies which cater for the interests of entomologists. These societies vary in their membership fees, aims and requirements for membership; they range from small local societies to international, professional organisations. It is a good idea to belong to one of these societies. Find out what societies exist where you live and go as a visitor to one or two of their meetings. From this you will be able to decide which one best suits your interests. In the major cities you may have quite a large choice; in smaller centres there may be only one society which caters for everyone interested in nature. In any case, there you will meet people of experience and knowledge who can help you get the most out of your collecting.

Some societies have excellent libraries, better than you can hope to accumulate yourself, and through them you may be able to borrow literature from more distant libraries through the inter-library loan scheme. Many societies publish their own magazine or periodical in which members publish their observations. In this way their knowledge becomes available to all.

As your experience grows your interests may change. If so, change your society. You will probably keep the friends you have made and make new ones. You will make your best contributions to knowledge and be happiest doing what interests you amongst people who have similar interests.

This book is intended as an introductory guide for those who want to make a collection of insects.

From making a collection to embarking on a more detailed study of insects is merely a step. If this book helps you to take the first step in that direction its aim will have been fulfilled.

ACKNOWLEDGMENTS

I should like to thank Howard Hughes, John Field and Heather McLennan of the staff of the Photography Department of the Australian Museum for advice and help in preparing the illustrations, and my wife for constructive criticism of the text, for typing the manuscript and preparing the index.

CONTENTS

CHAPTER 1

THE INSECT—STRUCTURE, FUNCTION AND DEVELOPMENT

IN ORDER TO UNDERSTAND how an insect lives and behaves it is necessary to understand something about its structural make-up. An insect's reactions to its surroundings may seem to us, with our human structure and senses, very strange. Some of the structures which insects possess are beautifully adapted to the tasks which they have to perform and some are almost unbelievably efficient at doing things which to us seem almost impossible. In fact, most of the fascination to be found in the study of insect life springs from the fact that insects are so wonderfully adapted to a world which is so different from our own. It should always be remembered that a 'specimen' was once a living, active being, and that each and every part of it once contributed to the life of that being. The 'specimen' is merely the remains of the body in which the most important 'part' of the insect, its life, was carried on. With an understanding of the insect's structure comes a deeper appreciation of the insect's beauty.

Despite the great wealth of different forms of insects which can be found, all are modifications of the same 'ground plan'. Remember that no one insect corresponds in all its features to the plan but that each species has more or fewer features which are modified from it. As this book is not intended to be a treatise on insect anatomy, only enough anatomical detail is given here to help in understanding later chapters. For those who wish to go into the matter in more detail, and this will be necessary for the more serious student, the works listed at the end of the book will indicate where further information can be obtained. At the same time, it would be useful to have on hand, as you read this chapter, a specimen of a large insect, such as a large grasshopper, cricket or cockroach and perhaps also a large beetle or butterfly, so that the various structures which are mentioned can be examined and some of the variety of forms in which one feature can occur can be appreciated. You may need a hand lens or magnifying glass to see some of them.

Most insects have a more or less cylindrical body, with the legs, wings and other appendages attached to it. The body is usually visibly made up of three sections: a fairly small head at the front, a middle section (thorax) to which are attached the legs and wings (though these are not always present) and a hind section (abdomen). Usually you can see that this is made up of a series of smaller ring-like segments. Each of these sections of the body is, broadly speaking, responsible for certain functions. The head, being at the front end of the body, is the first to come into contact with the surroundings of the insect as it moves about and it is, therefore, quite logically, that part of the insect which carries the main sense organs by means of which the insect receives information about the nature of its surroundings. It also carries the mouth-parts, around the mouth itself, by means of which food is taken up. The thorax is essentially responsible for locomotion; attached to it are the legs and wings. The abdomen contains mainly the organs responsible for digestion and assimilation of food, the storage of food reserves, the

Long horned grasshopper. A typical insect. The capsular head bears antennae, eyes and mouthparts, the thorax bears legs and wings and the cylindrical abdomen contains reproductive, digestive and excretory systems.

elimination of waste products produced through the various functions of the body and the organs responsible for reproduction.

THE HEAD

The head is usually in the form of a relatively small, hard capsule attached to the thorax by means of a membranous neck. It carries various organs, of which the most conspicuous are the antennae (feelers).

Antennae

In its most usual form each antenna consists of a long highly mobile filament, made up of segments hinged on one another by membranous joints; the flexibility of these joints accounts for the mobility of the antenna. The antennae are delicate organs of touch and they are the organs by means of which the insect detects scents. In those insects which depend greatly on their sense of touch, such as crickets and cockroaches, they are long. When the insect is not so dependent on touch they may be shorter, as, for example, in the dragonflies which hunt their prey on the wing by sight. In its simplest form each antenna consists of a series of cylindrical segments arranged end to end, but many species have this simple form modified; in cockchafer beetles for example, the last few segments are flattened into leaf-like plates which come to lie against one another in a group. In butterflies some of the segments near the apex are thickened to form a small club; in ants one segment is longer than the others with the result that the antenna as a whole appears to be 'elbowed'. In many insects the antennae are quite reduced in size and they can be difficult to find, whereas blind species which live in the darkness of caves often have especially long and sensitive antennae. The antennae bear large numbers of tiny sensory organs which are sensitive to chemicals in the air, in other words, they are scent-perceiving organs. This is a very important sense in insect life as insects often find their food, seek out

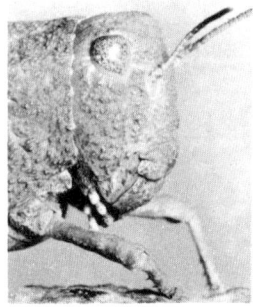

Insect head, the three pairs of mouth parts are at the lower end of the head.

their mates and select the plants on which to lay eggs by scent. In the Emperor moths the winged male can detect the scent of the female from a considerable distance away and fly to her. In species in which such a delicate sense is involved, the antennae are often 'feathery' or 'branched' to give a larger surface area for carrying the actual scent-receiving organs. Some of these antennae are extremely beautiful objects when seen under the microscope or magnifying glass. Flower-seeking insects are often attracted by the flower scents as well as by their colours. The droppings of animals attract flies within a very short time, the flies being very sensitive to the smells emitted by such substances.

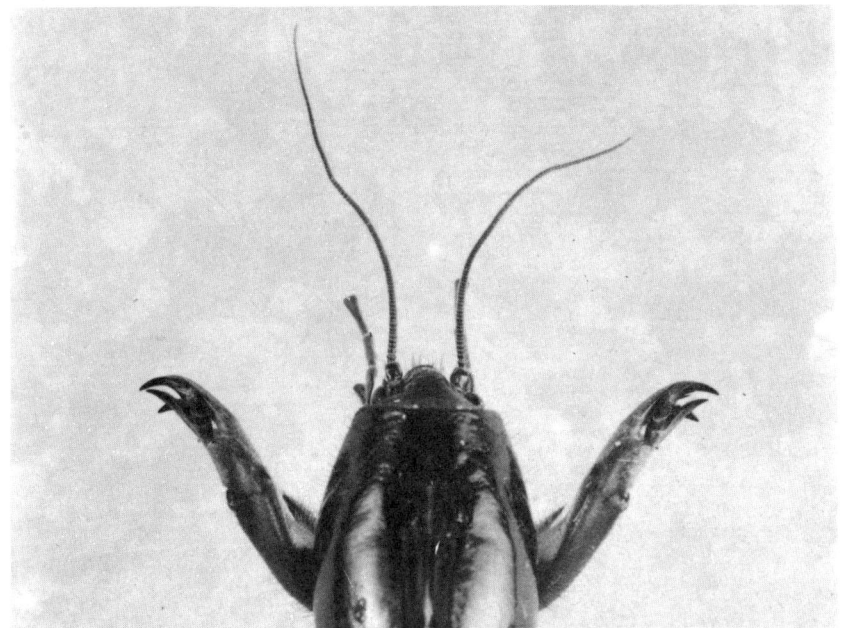

The simple, fine antennae of a mole cricket, very sensitive to touch.

'Lamellate' antennae of a scarab beetle.

11

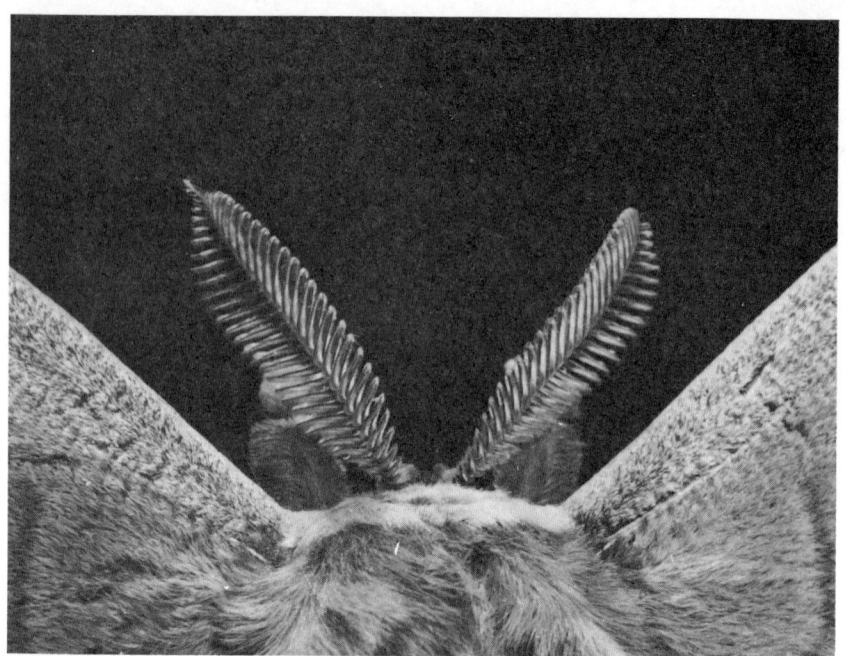

Complex antennae of an emperor moth, sensitive to odours produced by the female.

Antennae of a Rhipicerid beetle.

Antennae of a soldier fly.

In addition to the antennae, the head bears a pair of compound eyes and three simple eyes (ocelli; singular, ocellus).

Compound eyes

The compound eyes frequently occupy a large proportion of the head; the area varies, roughly, with the powers of sight possessed by the insect. Each eye is made up of a number of separate units (ommatidia), each ommatidium consisting of an outer lens, which appears on the surface of the eye as a single facet, and a complex inner structure receptive to light. The number of facets varies. In some almost-blind species there may be only a few, whereas in active predacious forms which seek their prey by sight there may be more than 25 000 in each eye, for example in some dragonflies. In the house fly there are about 4000 facets per eye; in butterflies the number may be as high as 20 000. If an insect eye is examined through a hand lens the mosaic formed by the facets can be seen. Each ommatidium receives light from a small part of the surroundings and the insect is thus sometimes said to have a 'mosaic' vision, that is, the image received by the eye is made up of a large number of separate points of light. As to the interpretation which the insect makes of this set of points we shall probably always remain ignorant because this interpretation takes place within the nervous system. This type of eye is well adapted to perceive movement of objects even though the lack of a focusing device in the eye may mean that the images formed are not the same as those formed by the human eye. The fact that insects are often associated with objects which appear to be coloured to us, flowers for example, suggests that we could expect them to be aware of colour themselves. Various experiments can be undertaken which show that some insects do discriminate between various colours, as opposed to merely distinguishing differences in the intensity of light. Bees seem to be able to distinguish four colours. They cannot distinguish between red, yellow or green but see these as one 'colour'; they see a blue-green and blue-violet and have the ability to distinguish between these and ultraviolet, to which, of course, our eyes are not sensitive and which we cannot, therefore, 'see'. The fact that they can distinguish red flowers from a background of green foliage is due to the fact that the red flowers concerned are also reflecting ultraviolet light. Many moths are attracted to ultraviolet light. There is little doubt that some insects do not distinguish separate colours but only appreciate a difference in light intensity; what such insects experience is even more difficult to imagine than in the case of colour-sensitive species. It is important to remember that although we can experiment to find out those things to which an insect is sensitive, we can never appreciate what the insect experiences, as we interpret our information in accordance with what we would feel and not with what is felt via an insect's nervous system.

The large, compound eyes of the predatory dragon fly occupy most of the head and are adapted for spotting moving prey.

Ocelli

The three ocelli are not present in all insects, and while they are clearly light-sensitive organs, their precise function is not fully understood. They are usually found on the upper surface of the head and each ocellus consists of a single lens with a light-sensitive structure below. Their structure suggests that they would be poor image-formers and some people regard them as organs which indicate to the insect only changes in light intensity.

Mouthparts

In addition to organs of sense the head also carries, on its underside around the mouth, the 'mouthparts'. These consist of three pairs of jaws. The first pair are the mandibles; these are usually hard, solid structures with a cutting edge and projections for grinding the food and they are used for biting and chewing. Behind the mandibles lie the maxillae—one maxilla behind each mandible. Each maxilla has two lobes at its free end, that is, the end furthest from the head; the outer lobe (galea) is usually fleshy and soft, and the inner (lacinia) is usually harder and spined or toothed on its inner surface. From the base of the galea on the outer side arises a short segmented appendage, the maxillary palp. This is sensory in function. The body of the maxilla assists in holding food. Behind the maxillae is the labium. This can be regarded as a second pair of maxillae which are fused together in the midline, with the parts corresponding to the galea and lacinia reduced in size. Labial palps corresponding to the maxillary palps are present. The labium forms the lower lip, as it were. A small flap (labrum) overlying the bases of the mandibles at the front of the head forms a sort of upper lip on under surface of which may be taste organs.

The mouthparts briefly described here are those typical of an insect which feeds on solid materials, such as leaves, bark, or other insects and are found in the cockroach and cricket as well as many other insects. These mouthparts are adapted for chewing but the multitude of different shapes in which the mouthparts do occur in correlation with the type of food taken by various insects is one of the finest examples of the adaptive variations on a basic theme which can be found in nature. In the sucking bugs (Hemiptera) which feed on plant juices the mandibles and maxillae are modified to form long, thin, sharp needle-like stylets and the labium is also elongated to form a trough in which the stylets lie. The stylets lie adjacent to one another to form canals through which feeding takes place. In mosquitoes (Culicidae)

The mouthparts of the cicada are modified to form a tube for sucking sap from plant tissues.

elongation of the mouthparts has occurred in a different way, and, including the labium, has resulted in a very efficient blood-sucking organ; in the larvae of lacewings the mandible and maxilla on each side form a fine tube through which the body juices of the prey are imbibed. In carnivorous insects the mandibles are sometimes very powerful and there are species in which the mouthparts are different in the two sexes. Some insects, mayflies for example, have the mouthparts so reduced so as to be non-functional. Such insects do not feed at all after becoming adult and may be fairly short-lived after maturing. Generally speaking, the mouthparts of a given species are well adapted to the food taken by that species and perusal of a textbook dealing with insect anatomy will reveal the wealth of variation to be found in these organs.

Neck

The neck, attaching the head to the thorax, is usually flexible. In some species the head is not very mobile, but in others, especially active predators, the head can be turned through a wide angle, as in the praying mantis, where the head can be swivelled loosely, giving a most alert and sometimes comical appearance to the insect.

THE THORAX

The thorax itself consists of a strong, usually rigid body, to which the legs and wings are attached. It is really made up of three separate segments, each equivalent to a segment of the abdomen, but in the case of the thorax the segments are usually rigidly fused to one another, and have lost their simple ring-like form. The first of the segments (prothorax) is usually small and bears only a pair of legs. The second (mesothorax) and the third (metathorax) each bear a pair of legs and also, in most adult insects, a pair of wings.

Legs

Correlated with the amount and the nature of the use to which they are put, the legs of insects vary tremendously in form. In fast-moving species they may be long and slender, as in the cockroaches (Blattidae). They may be broad and adapted for digging as in the mole crickets (Gryllotalpidae) or they may be reduced to mere props in species which use them only for alighting. In some cases they serve merely to suspend the insect from some support as in the crane flies (Tipulidae). Swimming insects often have fringed legs which function very efficiently as oars. In some grubs they are absent altogether. Mantids have front legs specially adapted for seizing prey. Each

The long slender legs of the native cockroach make it a fast runner.

The front legs of the mole cricket are adapted for burrowing in the ground.

The front legs of the mantid are adapted mainly for seizing and holding prey.

leg consists of several segments; there is the coxa, by which the leg is attached to the thorax, followed by the small trochanter which is usually attached rigidly to the next section, the femur. In jumping species, e.g. crickets, the femur may be thickened to accommodate the necessary muscles. Beyond the femur is the long, thinner tibia to which is attached the tarsus. The tarsus itself consists of segments, often five in number, the last of which bears the claws and may or may not have additional devices to assist the insect in maintaining a foothold.

Wings

The wings are important structures, although not all insects possess them, and they often show characteristics by which insects can be classified and identified. Each wing consists of a flat outgrowth from the upper part of the thorax hinged to the body of the thorax in a complex manner. This membranous flap is supported by the 'veins', a network of thickenings running through the membrane through which may run tracheae, tubes

16

forming part of the respiratory system, nerves and cavities containing blood. We shall return to the respiratory system later on. The arrangement of the veins in the wing is not haphazard but conforms to a pattern which is laid down largely around tracheae during the development of the wing. The various veins have been named and a basic pattern can be recognised in most insects although it varies considerably from one group of insects to another. In many groups of insects a special terminology is used for the veins, so that several sets of names are in current use according to the group of insects concerned. These various terminologies were invented independently for each group before it was realised that the patterns of each could usually be referred to a basic pattern. This fundamental pattern, with the names of the veins and the many variations and arrangements in which they can occur, is described in textbooks of entomology. It is essential that the serious student consult one of these. This basic pattern is altered by addition, reduction, fusion and additional forking of the veins and by the addition or loss of cross-veins. The relative positions and strength of the veins also vary from one group of insects to another in ways characteristic for each group.

Winged insects usually have four wings and it is normal for all four wings to be used in flight. Generally, in more advanced insects such as bees, butterflies or moths the wings of each side are coupled together by some device so that they beat as one. In some four-winged groups they beat independently of one another, the dragonflies for example. Most insects are capable of flexing the wings over the back when they are not in use, but this is not true of dragonflies or mayflies. Quite often the fore wings are modified to form a protective shield for the hind wings and may not be used in flight. In its extreme form this is seen in the beetles where the fore wings are modified to form hard 'elytra' under which the membranous hind wings are folded when at rest. In grasshoppers and locusts the fore wings are thickened, but not to the extent of beetle elytra, and the hind wings are folded fan-wise under them when the insect is at rest.

Although the winged insects usually have four wings, many groups have only two. The 'true' flies (houseflies and bushflies, e.g.) have lost the hind pair, which is now represented by a pair of knobbed organs, the halteres, and many insects, in some cases whole groups of species, have lost the powers of flight altogether. This occurred in the case of fleas and lice. Amongst almost any group of species of normally active flyers can be found some which are flightless. While most beetles can fly, species of a few families cannot and in some families which can there are sometimes some species which do not. This is true of the ground beetles. Some species of sucking bugs cannot fly whereas other members of the same families can. In some species one sex may be winged but the other not. It may be mentioned here that flight is an attribute of adult insects only; immature specimens cannot fly.

The power of flight is shared by very few groups of animals. True flight is found today only in birds, bats and insects. In insects it has been responsible for much of the success of the group: it enables them to colonise new areas, seek out food and mates and assists them to flee from enemies. Insects, apart from their active flight, are frequently transported tremendous distances passively by air currents and specimens have been found far out to sea and high up in the atmosphere.

THE ABDOMEN

The insect abdomen is usually made up of eleven segments although in many insects there are fewer due to reduction, fusion of segments with one another, or other modification having taken place. Basically, each segment is ring-like and joined to the segment before and behind by a thinner membrane. In practice, however, the segments are modified from this simple plan. The segments are movable, one on the other, and the abdomen as a whole therefore made flexible. The segments may also telescope into one another. In the remote ancestors of insects the abdominal segments probably carried limbs, as did the thoracic segments, but in most of the insects found today these are no longer present, although the ovipositor (or egg-laying organ) of female insects is believed to be made up of the highly modified remnants of parts of the legs of the eighth and ninth abdominal segments. In some insects the ovipositor is very conspicuous, as in katydids, and in some parasitic wasps where it protrudes well beyond the hind-end of the body. In others it is not so obvious and in bees, ants and wasps it has been converted to form the sting. It is interesting to note that only females sting.

Near the apex of the abdomen in the male are to be found the male external genitalia. These vary greatly from group to group and consist of a variety of organs, often including claspers of diverse origins used during mating. These external genital organs of both sexes, (claspers etc. of the male and ovipositor and other organs of the female) have been found to be very useful structures for differentiating between species in many insect groups. It has been found quite frequently that two species which are virtually indistinguishable in other ways can be recognised after a study of their genitalia has been made.

At the end of the abdomen lies the anus, the posterior opening of the digestive tract, and adjacent to this is usually a pair of small segmented organs, the cerci, which have the appearance of small projections and which function in much the same way as the antennae do. In a few insects they are quite long as in mayflies, but generally they are fairly short. They are sometimes altered in form to carry out special functions as in the case of the forceps at the end of the abdomen of earwigs.

THE CUTICLE

So far we have dealt, in the broadest outline, with the external features of the insect's morphology. Before going on to discuss briefly the internal organs and their functions, mention must be made of the insect skeleton. This is not an internal structure as it is in humans or the other vertebrate animals—that is, other mammals, fish, birds, reptiles or amphibians. The insect skeleton, or cuticle, is also the outer covering. It is made up of complex chemical substances, amongst which and of special importance is chitin. This is a hard, impermeable, light, strong material. The cuticle of the insect varies in thickness and hardness. For instance, the segments of the abdomen are hard and rigid whereas the membrane between them is soft and flexible. The thorax is mostly especially hard and rigid but is somewhat flexible along certain defined lines which allow distortion of the thorax by muscular action during movement and flight. The joints of the legs are similarly softer, membranous areas of the cuticle between the harder rigid segments of the leg. At certain points within the cuticular skeleton, the muscles of the body are attached which effect the movement of the various parts of the body. The cuticle is frequently infolded to form ridges which serve for the attachment of muscles.

SOUND AND HEARING

The insect's body usually carries hairs and spines of various kinds and sizes. Many of these hairs are served by nerves and are sensitive to touch, with the result that the insect is made aware of the presence of near objects. In some insects these hairs respond to vibrations in the air, in other words, they form a simple type of ear in that they inform the insect of air vibrations. These hairs are found on some caterpillars, which can be stimulated into action by making a noise of appropriate pitch. Other insects, such as the cicadas, crickets and many moths have special, complex organs by means of which they hear. In the case of locusts, cicadas and moths these organs are on the abdomen; in the case of the crickets and katydids they are on the tibiae of the front legs. Many of the insects with more complex hearing apparatus belong to species which are themselves able to make sounds. The crickets and their relatives do so by means of a special scraping apparatus at the bases of their wings; locusts and grasshoppers make a sound by rubbing the hind legs against the thickened fore wings and cicadas by means of a complex and most efficient sound producing organ at the base of the abdomen. The range of pitch appreciated by insects does not correspond to our own and in many cases it seems likely that they appreciate the rhythm of the sounds rather than the actual tones.

INTERNAL ANATOMY

Digestive system

One of the most important activities undertaken by insects is that of feeding and we have already mentioned how the mouthparts responsible for actually taking up the food are modified to suit the type of food taken. Feeding involves more than merely eating. The food must be broken down, digested and taken into the body so that it can be distributed by the transporting blood to the organs which require it. The alimentary canal extends from mouth to anus; it usually has at least a few convolutions which increase its length and hence its absorptive area. The front part (fore intestine) is lined with a thin cuticle and usually consists of a narrow tube, the oesophagus, which expands into a crop in which food is stored. Behind this may be a strong muscular gizzard, the walls of which are hardened and roughened. Behind the fore intestine is the mid intestine. The minute anatomy of this section is different from that of the fore intestine and it varies considerably in length. It is here, in the mid intestine, that the digestion and absorption of the food takes place. The surface area of this part of the intestine is frequently increased by 'branching' and 'coiling'. Behind the mid intestine is the hind intestine which, like the fore intestine is lined with cuticle. Here water is re-absorbed into the body from the gut contents. The insect body is small, and therefore the surface area of the body is relatively greater per volume than it is in a larger animal. If the insect were to lose water by evaporation from this relatively large surface, as we do in perspiration, it would soon lose more than it could replace and die of desiccation. For this reason it is important that the insect cuticle be impermeable and that as much water be retained in the body as possible and not lost together with the waste products of the body. Near the junction of the hind and mid intestine the important malpighian tubes enter the hind intestine. These are the organs which are responsible for the removal of waste substances from the blood and their ejection from the body; the waste substances from the blood pass down these tubes to the alimentary canal and are ejected, after water has been removed from them, with the solid wastes.

In addition to the alimentary canal, insects possess salivary glands which open near the mouth and which produce substances undertaking various digestive functions. In some predacious species, for example the larvae of lacewings, the products of these glands are injected into the prey where they

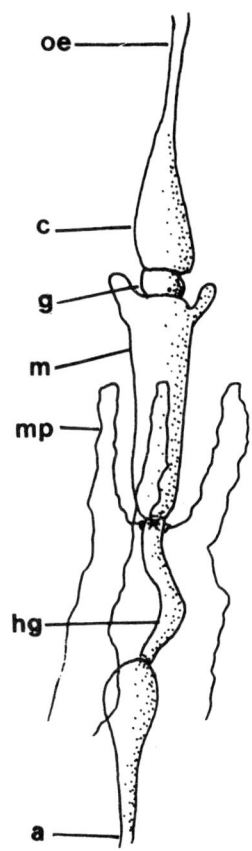

A simple form of digestive system: oe *oesophagus;* c *crop;* g *gizzard;* m *midgut;* mp *malpighian tubes (excretory);* hg *hindgut;* a *anus.*

19

soften the tissues and so enable the body contents of the prey to be sucked up. Silk, which is produced by many insects, is usually a product of modified salivary glands. Whereas most insects feed on substances which are fairly easily digested, such as those containing carbohydrates, fats and proteins as in our own diets, some are capable of feeding on unusual substances. Thus, termites feed on wood, clothes moth larvae feed on wool and the wax moth, which is a pest in bee hives, feeds on wax. In the case of termites and the wax moth digestion of their food is assisted by bacteria and other minute living organisms which live in their digestive systems.

Respiratory system

Breathing is the action by which oxygen is brought into the body to take part in the chemical processes which release the energy used in movement and the other physiological processes. Oxygen only beomes available to an animal after being taken up on a moist surface. In the case of humans it is taken up on the moist surface of our lungs and in the process of breathing out we lose considerable quantities of water. As we have mentioned, water loss is a great problem to a small animal such as an insect and this difficulty has been overcome by the use of a completely different respiratory system. Insects breathe by means of a system of fine tubes (tracheae) which extend throughout the body, branching and becoming finer and finer until the finest branches (tracheoles) are in actual contact with the organs requiring the oxygen. In this way, oxygen is taken directly to the organs and not carried there from lungs by the blood as in humans. An examination of the side of an insect abdomen will reveal the presence of a row of tiny holes; these are the spiracles, outlets and inlets through which the air gains access to the tracheal system. In some of the heavier, faster flying insects the tracheae are expanded at certain points to form large air sacs; these are quite extensive in some of the larger beetles and in such insects as the bee. The spiracles usually have an apparatus whereby they can be opened or closed to alter the air flow. In addition to the inhaling and exhaling of air due to ordinary body movements and the natural diffusion of air through the system many insects exert a pumping action by contraction and expansion of the abdomen. If a bee is caught in active flight the abdomen will be seen to contract and expand for some time after active flight activity, which required much oxygen, has ceased.

The immature forms of many insects live in water and these may have gills which usually consist of a leaf-like expansion of some part of the body (usually on the sides of the abdomen) which contain many tracheae. Oxygen from the water is taken into the tracheae via the gills which are usually lost when the insect becomes adult and leaves the water for life on land.

Circulatory system

The circulatory system of insects is comparatively simple in so far as the blood merely bathes the organs of the body and is not transported to and from them by a complex system of arteries and veins. There is a single tubular heart running along the dorsal side of the insect. The blood is drawn into it through valves along its length and pumped forward towards the brain. From there the blood flows freely through the body and appendages. It receives nutrients from the alimentary canal and these are passed on to the various organs of the body and from these, waste products are carried by the blood to the malpighian tubes. The physiological processes involved in this are, of course, extremely complex and a considerable amount of energy is used in carrying them out. The blood also carries various glandular products, the hormones, which set in motion or control many of the body's activities, such as the periodic moulting of the cuticle.

Nervous system

The central nervous system, through which the activities of the insect are coordinated, consists of a brain which lies above the oesophagus in the head and a ventral nerve cord which lies along the body below the alimentary canal. Behind this the main nerve cord, extending backward along the body, gives off nerves to the various organs.

Reproductive system

The adult insect has the responsibility of ensuring that the species remains in existence and in most insect species both males and females are known. In the males the internal reproductive organs consist of a large pair of testes which

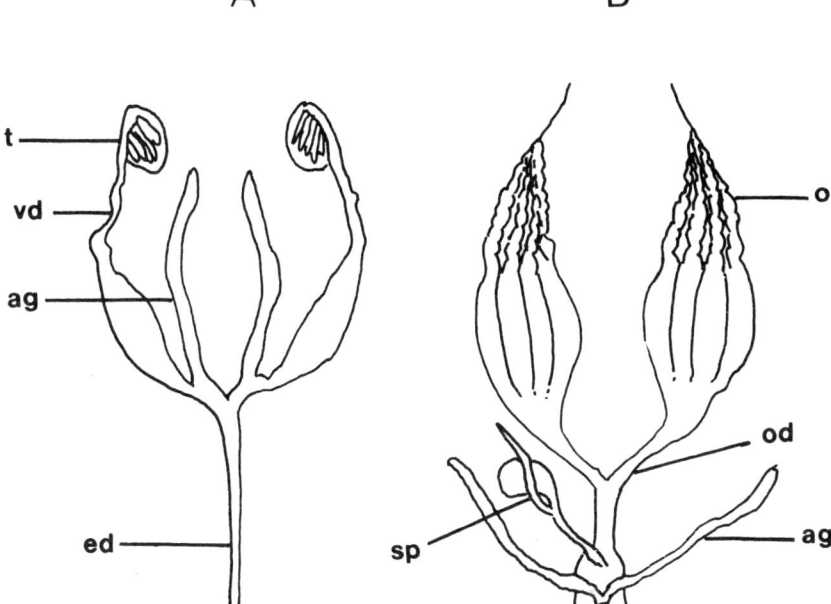

Simple reproductive systems. Male (A): t testes; vd vas deferens (duct through which sperm passes); ag accessory gland which secretes substances needed by sperm; ed ejaculatory duct to exterior. Female (B): o ovary; od oviduct through which eggs pass; ag accessory glands; sp spermatheca in which sperm from male is stored.

lie in the abdomen and in which sperm are produced. In many species the sperm are enclosed in a case, the spermatophore, produced by special glands and it is the whole spermatophore which is transferred to the female. The female internal reproductive organs consist of a pair of ovaries, additional glands and a spermatheca which open onto the oviduct, the spermatheca is a storage chamber, for sperm previously received from the male. The eggs are fertilized after the egg shell (chorion) has been laid down around them just before they are laid.

MATING

In insects the sexes often meet purely by chance or because they happen to be attracted to the same specific habitat such as a particular type of tree or flower. In other cases meeting of the sexes is ensured by their being in flight for a short time only, as in the case of ants. Here the winged males and females leave the nests under certain conditions and as many are about at the same time the chances of meeting and mating are good. Many male moths are attracted to their females by scent and this is probably true for many other insects, particularly once they are in proximity to one another. Mating may be a momentary affair, as in some wasps, or the copulating pair may remain attached to one another for a considerable period, even a matter of days. There is frequently a prenuptial dance, of varying complexity, sometimes undertaken by one sex only and sometimes by both. There are species known, such as the scorpion fly, in which the male offers food to the female and whilst she is feeding mating takes place. In many species the male has been dispensed with, the female producing eggs which mature without fertilisation, a phenomenon known as parthenogenesis. This is known in many groups of insects, especially in the aphids, in which generation after generation may be produced rapidly under favourable conditions without the intervention of males.

Not all insects lay eggs. Some, for example the flesh fly and many of the aphids, give rise to living young, a phenomenon called viviparity.

One of the reasons for the success of insects in nature is their ability to reproduce rapidly and to produce large numbers of offspring. As a general rule species which are exposed to great hazards and which suffer high mortality rates produce more young than species which are not subject to such losses. Parasitic species in which a host has to be found by search may lay large numbers of eggs, perhaps many thousands in a batch.

DEVELOPMENT AND METAMORPHOSIS

Eggs

Most insects lay eggs, the form and size of which vary enormously from group to group. Most of them are small, many are beautifully sculptured and some are of peculiar shapes. Some are hard-shelled, especially those laid in exposed situations. Those which are laid within plant tissues or in other damp, protected situations are usually soft-shelled. Insects' eggs may be laid in batches or they may be dropped singly, all at once or over a long period of time, haphazardly or after great pains have been taken to seek out a suitable site—all depending on the species concerned. Almost every possibility is found realised in one or other species.

Growth

An insect is not born in a mature state; it must grow and develop from youth to old age. Most insects start life as an egg and the young, on leaving it, feed and grow. Growth in the insect is an irregular affair due to the fact that the external cuticle is comparatively rigid and has only very limited stretching powers. The cuticle inself is laid down by a layer of tissue, the epidermis, which lies below it and after a period of feeding it becomes necessary for the insect to moult, casting off the old cuticle. This process is very complex and is under the control of hormones secreted by special groups of glandular cells within the body. When an insect is due to moult it becomes inactive, a new cuticle is laid down below the old one by the epidermis, the old cuticle is partially dissolved away and eventually, by its own efforts, the insect splits the old cuticle, usually along definite lines of weakness on the head and thorax, and emerges from it. At this stage the new cuticle is soft but it hardens with time, becomes rigid and takes on the colouration proper to it. It is very commonly the case that an insect's appearance differs after a moult because the new cuticle being laid down may have structures, such as knobs or bristles, or colour patterns, differing from those of the previous cuticle. The period between each moult (ecdysis) and the next is called a 'stadium'

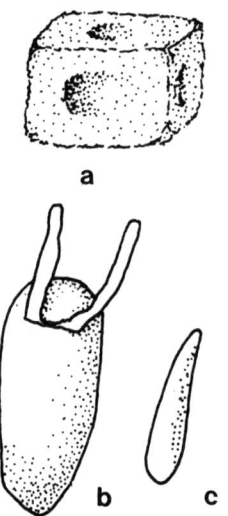

Various forms of insect eggs
a *scorpionfly*
b *vinegar fly*
c *house fly*
d *beetle*

and the form adopted during that period is known as the 'instar'. The number of stadia and instars through which an insect passes from egg to adult varies considerably. Cockroaches have about seven moults, some of the mayflies may moult as many as thirty times whilst most caterpillars moult less than eight or nine times. It is usual for the number of moults to be constant for a species but exceptions do occur from time to time. It will be seen that this arrangement allows very different forms to occur during the life span of a single insect and the potentialities of this have been exploited to the full.

Types of metamorphosis

In the more primitive orders of insects there is comparatively little change in form from the time the young leave the egg until the insect is adult; in fact, the main changes are in size and sexual maturity and there may be even some moults after maturity. Insects which have this simple type of development include the bristle-tails, silver fish and spring-tails.

In the somewhat more advanced groups (sometimes called Exopterygota) the young insect leaves the egg in a form which more or less resembles the parent but differing, of course, in size and in lacking wings. There is no pupa (or chrysalis) in the life cycle of these insects and in the young form they are referred to as nymphs. As the nymph grows and moults wing rudiments appear on the thorax; these become larger with each moult, reaching their fully developed and functional condition at the final moult whereupon the insect is of adult form although it may not be capable of reproduction until later. In such insects the food and mode of life of the nymph is usually similar to that of the adult. This gradual change from young nymph to adult is referred to as hemimetabolous, direct or incomplete metamorphosis. Insects with this type of development include mayflies, dragonflies, damselflies, stoneflies, grasshoppers, locusts, crickets, stick- and leaf-insects, earwigs, cockroaches, mantids, termites, psocids, lice, plant bugs, other sucking bugs and thrips.

In the higher groups of insects (sometimes called Endopterygota) the young are very different in appearance from the adults. They hatch from the egg as a larva. There are many forms in which larvae occur and modification to suit particular environments results in some of the forms being peculiar to a group of species. Larvae may take the form of caterpillar, grub or maggot.

During the course of development of some species, for example, oil beetles, the larva is of a different type in different instars. This change in form is known as hypermetamorphosis.

After growing and undergoing several larval moults a final larval moult is undertaken which reveals the immobile pupa (chrysalis). During the pupal period activity and feeding usually cease; in many cases this period is spent in a specially constructed cocoon or earthen cell and the body structures undergo considerable change within the pupa. This stage is often referred to as a resting stage but this is not so as, although the pupa may be quiescent and still great internal changes are taking place as the body of the adult is laid down within the pupal skin.

When the pupa finally moults the insect emerges in its adult form. This type of metamorphosis, involving egg, larval and pupal stages, is known as holometabolous, indirect or complete. The holometabolous type of development allows one species of insect to become specialised for more than one way of life. The larva is essentially a feeding and growing stage and its structure is extremely specialised for the efficient carrying out of these functions. The pupal stage allows this form to undergo the considerable

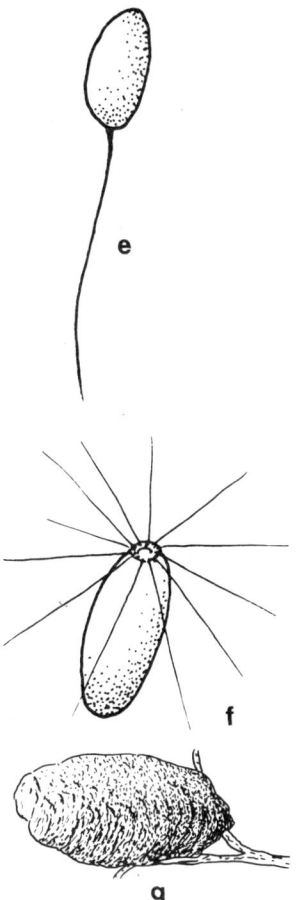

e *lacewing*
f *water scorpion*
g *mantid egg case,*
containing many eggs

Butterfly's egg.

23

Caterpillar—the immature, growing stage in the butterfly's life cycle.

The adult butterfly, the result of a 'complete' type of life cycle.

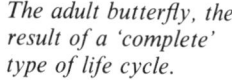

Pupa (chrysalis), the life cycle stage in which the body of the butterfly is built up.

changes required to produce the very different adult, which is essentially a reproductive and dispersive phase of the life cycle. Insects with this type of development include the lacewings, antlions, scorpionflies, butterflies, moths, caddisflies, true flies, fleas, ants, bees, wasps, ichneumonflies and beetles.

POLYMORPHISM

In most species the only differences between adult members of the population are differences due to sex; this difference is referred to as sexual dimorphism. In some groups, however, it is found that within a single species individuals may occur in various forms. This state of affairs is referred to as polymorphism. For example, some species have winged males and females but individuals of both sexes may be found without wings in the adult state. In the ants, there are winged male and female individuals which are responsible for reproduction and there are adult females without wings which have degenerate reproductive organs and which spend their lives attending to the needs of the colony. The degree of differences which may be found varies considerably from the extreme differences as found in the ants to species of other groups in which the forms differ only in minor details.

CHAPTER 2

NAMING AND CLASSIFICATION

ZOOLOGISTS HAVE GROUPED THE vast number of existing animal species known into a number of groups (phyla) and the members of each phylum, although they may superficially appear to be very different and may live in very different environments, share a common basic plan of anatomical structure. The phylum to which the insects are assigned is known as the Arthropoda, a name which means 'joint-legged'. Included in this phylum are also such creatures as lobsters, shrimps, crayfish, barnacles, spiders, mites, ticks, scorpions, centipedes and millipedes as well as many less well known animals both living and extinct. The features which all these animals share and which are considered as characteristic for the phylum are, broadly, as follows: the body is divided into a number of segments from at least some of which arise jointed appendages. The body segments are grouped together to form distinct regions of the body (e.g. head, thorax etc.) and the appendages are specialised in different sections of the body to perform various functions. The cuticle is hardened in special areas to form a casing to which the muscles are attached. The heart is dorsal in position and the circulatory system is open, that is the blood bathes the organs and is not carried in a complex system of arteries and veins.

The Arthropoda have been divided into a number of classes, and each of these classes exhibits certain special features in addition to the characteristics of the phylum as a whole.

The insects have the segments of the body grouped so as to form a head, thorax and abdomen. Antennae are present and there are three pairs of jaws. There are three pairs of walking legs and, in the adult, usually two pairs of wings although loss by reduction is common and some orders have evolved from wingless ancestors. There are no abdominal legs corresponding to those of the thorax. Immature forms may be very different from the adults; metamorphosis of varying complexity is usual.

It has been found that there exist in nature an enormous number of kinds or species which fall within the definition given for the class Insecta. Of all the species of animals known probably about 75 per cent are insects, that is, there are many more species of insects than there are of all other kinds of animals put together. In fact, it is impossible to say how many different species of insects have been found and described by entomologists. A conservative estimate would be about 700 000 species, and thousands more are described annually! Quite clearly it is impossible for one person to be able to recognise all, or even a fraction, of these species and so we find that it is necessary for each person to limit his or her detailed studies to a relatively small number of species. Needless to say, there is no single collection of insects in the world which contains even a single specimen of each known species.

Scientific names

If we wish to speak or write about something it is essential that it has a name. We have invented a great many words to serve as names for the objects we see and use around us in the world. It is also necessary to name the many species of insects which are known so that reference may be made to them. If we tried to use a different name for each species we would soon be searching for suitable words.

Zoologists, whether studying insects or any other group, have come to use a standardised method of naming species which is both simple and effective and has the advantage of being completely international and understandable by all. Many people are repelled by the 'Latin' or scientific names given to animals and they prefer to use the so-called 'common names'. However, the advantages of Latin names are immediately apparent to anyone trying to obtain information from entomological literature, moving out of their own country or, in many cases, even out of their own circle of friends. Many of the 'common names' are not common at all, in any sense, and are used by a very limited number of people. Latin words are not really any more difficult to use than those of any other language and we can use these words as names without knowing their meaning, as we very often do in our own language.

The reason for the scientific names being in Latin is partly historical, but it has been found to be very practical to continue the tradition. At the time when this system of using two Latin words was adopted generally (in the mid-eighteenth century) Latin was the language of all learned writings; thus it came about that the descriptions and scientific names of the animals were in Latin.

For any entomological discussion to be intelligible or for information to be passed accurately from person to person it is absolutely necessary to be sure that everyone understands to which species reference is being made. The butterfly which bears the scientific name *Danaus plexippus* is called the Wanderer in Australia and the Milkweed Butterfly or Monarch in America. It is also known as the Black-veined Brown as well as by several other names. No matter how many local names may be used for a species, the scientific name refers always to only one species no matter where you are. Scientific names are never intentionally duplicated.

The scientific name of each species always consists of two Latin or Latinised words. This system of using two Latin words for the insect's name is referred to as the binomial or binominal system. The name is usually printed in *italics* or other distinctive type, or when written or typed is underlined. The first of the two Latin words of the name (the generic name) is always written with a capital letter but the second (the specific name) is never written with a capital, even when it is derived from the name of a person or a place as in *Zorotypus hubbardi* or *Silvestricampa africana*.

There is now more or less general agreement on how scientific names should be formed, their grammar, derivation and other kindred matters. These are covered by international rules, which have been published in a document called the 'International Code of Zoological Nomenclature'. The object of these rules is to obtain as much stability and consistency in animal names as possible, while at the same time allowing for changes of names made necessary by increased knowledge or discovery of errors. In general, if a species which has been described and named is later described again and renamed unknowingly the first name stands and the later name is referred to as a 'synonym'.

No two animal species can have the same scientific name. Occasionally it happens that two species are described by different people but are given the same generic and specific name. In such a case, when the error is noticed, the later-described species must be given a new name, the original name being retained for the first-described species. A generic name can never be repeated; specific names can be repeated but must not be used for more than one species in a given genus. Thus we can have *Metapa natalensis* and

Basutacris natalensis as the names of two species in different genera but there cannot be two species called by either of these name combinations.

In addition to laying down rules concerning the formation and application of scientific names for species, the International Code also contains rules concerning the formation and application of generic and family names and indicates what is required in a description before the name which is being used for the species concerned can be considered valid.

The rules are, in some ways, fairly complex but this is due to the fact that they are an attempt to cover all the possibilities which can arise; in practice they show quite clearly which names would be applicable and which not in any given case.

Generic names

The use of two Latin words for the scientific name, instead of one word, makes it possible to use the name as more than a mere label. It is possible to use it to indicate the relationship of different species to each other. *Danaus plexippus* is a large brown species with a complex colour pattern including black colouring along the wing veins and a black band with white spots along the edge of the wing. In America there are several other species of butterflies which resemble *D. plexippus* in anatomy and colour but which differ in various details of body structure and patterning. These are all distinct species and each, of course, has its own scientific name. They include such species as *Danaus gilippus*, *Danaus erisimus* and *Danaus cleophile*. As will be seen, these three species have the same generic (first) name as *Danaus plexippus* and this indicates that they are closely related to *D. plexippus*. They are said to belong to the same 'genus'. From this it will be seen that a genus consists of species which are closely related to each other and this is indicated in their names by all having the same generic name. Each species in the genus has its own distinctive specific name which applies to it only and to no other species of the genus.

Sometimes a person's name will be mentioned after the scientific name. This is the name of the person who first described the species. Species names will frequently be found in which the name of the author is placed in brackets after the Latin name of the insect. This indicates that the species was placed in a certain genus by the original author but that, for some reason, it was subsequently found necessary to include it in a different genus. For example, the psocid *Philotarsus greyi* Edwards was found, many years after its description, to be more closely related to species of the genus *Haplophallus*. It was, therefore, transferred to the latter genus and is now referred to as *Haplophallus greyi* (Edwards).

Euploea is a large genus of butterflies which in many of their characteristics resemble species of the genus *Danaus*. They have, however, as a group, some distinct differences from the species of *Danaus* and are considered to be less closely related to the species included in *Danaus* than they are to each other and so we have two separate genera (the plural of genus) which, despite their differences, have some characteristics in common.

Genera which are related to one another are grouped into a family.

Family names

Each family of insects is given a name derived from the name of one of the genera in it, the genera *Euploea*, *Danaus* and many others not mentioned here, are included in a family, the Danaidae. A family name is always derived from the stem of a generic name by adding the suffix -idae. Thus, all animal family names end in -idae. In print, family names are written with a capital letter and are never italicised or underlined.

There are a great many families of insects; many of these have a characteristic facies, or appearance, and their members can be recognised as belonging to them even though it may require a specialist to name individual species. For example, the great family Curculionidae includes the weevils. A weevil is easily recognised as such although there is probably no single person who can identify all of the estimated 40 000 known species. This is the largest family in the animal kingdom. The Vespidae are wasps, and here again many of them are easily identified as belonging to that family on sight. In fact, each group, whether it be genus or family has had its characters defined in the entomological literature. We thus have a hierarchy of defined groups: species are grouped into genera and genera into families.

Groups above family level

The matter does not end there, for related families are grouped into 'orders'. An order is thus a group of families having many characters in common. Opinion as to the number of orders into which the insect families should be grouped differs from one authority to another and changes have been made from time to time as knowledge of the insects themselves has increased. The same applies to the grouping of genera into families and species into genera. There is no accepted suffix by which the names of the orders can be recognised as such but many of the insect orders end in the letters -ptera. In this book we shall group the insects into 30 orders.

The insect orders are themselves grouped to indicate those which have features in common with one another; this grouping is given at the end of this chapter, in the synopsis of orders, with an indication of the main characteristics of each group and each order.

Members of the various orders are often easily recognised as such and many of the orders have overall common names which cover all members in a general way. For example, the beetles belong to the order Coleoptera, the mayflies to the Ephemeroptera, the dragonflies to the Odonata, the butterflies and moths to the Lepidoptera, the caddisflies to the Trichoptera and the scorpionflies to the Mecoptera.

It will be seen from the common names used for many insects that the word 'fly' has little value in indicating relationships and that the word forms part of the English common names of many insects which are only remotely related. The so-called two-winged or 'true' flies belong to the order Diptera.

Other categories

In addition to the categories of genus, family and order it is often useful to use divisions intermediate between these. For these the prefixes sub- and super- are often used. There are thus subfamilies (groups of genera within the family) with names ending in -inae, superfamilies (groups of families) with names ending in -oidea and suborders (groups of superfamilies within the orders) with name endings not standardised. By use of these categories it is sometimes possible to show in detail the suspected relationships of the insects

If we wish to indicate fully the position occupied by say, the Monarch butterfly in one of the presently accepted classifications of the animal kingdom we can do so as follows:

Phylum: Arthropoda

Class: Insecta

Subclass: Pterygota

Division: Neoptera

Order: Lepidoptera

Suborder: Ditrysia

Superfamily: Papilionoidea

| Family: Danaidae | Genus: *Danaus* |
| Subfamily: Danainae· | Species: *plexippus* Linnaeus |

In practice it very seldom happens that it is necessary to trace out the full position of a species in this way as the relationships of the higher categories are usually well known and it is usually generally known, for example, to which order a particular family belongs. If not, such information can easily be obtained from textbooks or other entomological literature.

Objectives of naming and classification

It is as well to remember that the naming of insect species and their classification into groups within groups serves two distinct purposes. First, we name the species so that they can be referred to without ambiguity, and it is highly desirable that this naming be stable. This is achieved in some measure by international agreement on how the names are to be formed and applied but changes must take place with increasing knowledge. Secondly, it so happens that a binomial system has been devised which allows relationships between species to be indicated at the generic level. It is but a step to extend this grouping into higher groupings to indicate the various degrees of relationship between the groups; that is, species within a genus are more closely related to each other than they are to species of other genera and still less closely related to species of other families and orders. It is impossible to obtain complete stability at present in this higher classification as we can only base our classification on present knowledge. As time goes on more species are discovered and more is learned about known species which enables us to alter our classification so that it more nearly reflects what is thought to be the true relationships of the species concerned. It is unlikely that we shall ever achieve a completely satisfactory classification, but we can continue to improve the presently accepted classification to the best of our ability and in accordance with knowledge of the insects themselves. From this it is clear that there is no such thing as a 'correct' classification. When classifications differ it is because those people who made them differ in their opinion as to the relative importance of the characteristics available for use as the basis for their groupings or because one classifier had more information on the insects than another had.

In any event, the naming and classification of insects is designed to arrange a vast number of entities, too many for us to think about individually, into some sort of system which will enable us to think about them—in fact, to bring order out of chaos. It is, therefore, a convenience. At the same time an effort is made to express the relationships between the species. It is not always possible to do both things satisfactorily in one classification and our classification is a compromise.

One final word: we must remember that the insects were here before we were; our classifications must depend on what we find in nature and with new facts to hand we must be prepared to alter our classifications. We have defined genera, families and orders, but we cannot expect the insects always to conform to our definitions and it is often the case that we have to alter our definitions when new facts come to light.

SYNOPSIS OF ORDERS OF INSECTS

In this synopsis the main characteristics of each insect order are set out, the orders being grouped into higher categories. An indication of where they are to be found is given as well as some of their common names. In Chapter 6 you will find a key to the orders, with the aid of which it should be possible to determine to which orders most typical specimens belong.

Subclass APTERYGOTA

Small wingless insects, considered to have been evolved from wingless ancestors. No true metamorphosis, the young and adults being very similar to one another except in size.

1. Order THYSANURA

Bristle-tails, Silverfish.

Antennae long. Two long cerci and a median segmented tail filament, giving a three-tailed appearance to the insect. Mouthparts formed for chewing. Tarsi two- to four-segmented.

Found in soil, in rotting leaf litter, under stones and in ant and termite nests.

2. Order DIPLURA

No common name.

Antennae long. Two long cerci, sometimes modified to form claspers; no median tail filament. Mouthparts sunken within head. Tarsi unsegmented. No compound eyes, no ocelli.

Found in soil, fallen leaves, in decaying wood or under stones.

3. Order PROTURA

No common name.

Very small insects. No antennae. No cerci. No compound eyes. No ocelli. Mouthparts of piercing type, sunken within the head capsule. Tarsi unsegmented.

Found in soil, under bark, in leaf litter and under stones.

4. Order COLLEMBOLA

Springtails.

Antennae may be long, but made up of few segments. Mouthparts formed for chewing, sunken within the head capsule. No compound eyes. Abdomen formed from only six segments. Abdomen with a ventral tube protruding below first segment; a forked organ used for springing on fourth segment and a small structure (retinaculum) on third segment for retaining the fork when not in use. These abdominal structures are found only in Collembola.

Found in soil, in leaf litter, under bark, on leaves, in ant and termite nests. Occasionally found in enormous numbers on the surface of puddles or pools.

Subclass PTERYGOTA

Winged in the adult stage; if wingless, considered to have evolved from winged ancestors. Metamorphosis hemimetabolous or holometabolous.

Division PALAEOPTERA

Metamorphosis hemimetabolous. Pupal stage normally absent. Developing wings visible externally on young stages (nymphs). Wings cannot be flexed over back.

5. Order EPHEMEROPTERA

Mayflies.

Antennae thin and short. Mouthparts reduced to non-functional remnants. Membranous wings held upwards when not in use, fore wings much larger than reduced hind wings. Venation complex. Cerci long with or without a median segmented tail filament. Nymphs aquatic, with long cerci and often with median segmented filament. Gills present along sides of abdomen. After emergence from last nymphal skin, the winged form undergoes a moult before maturity. (This is the only insect group in which the winged form is known to moult; winged form is known as a subimago before its moult to the true adult or imago).

Found alongside the streams, lakes, ponds etc. in which the nymphs develop. Sometimes very short-lived. Found in dense swarms over water.

6. Order ODONATA
Dragonflies, damselflies.

Antennae small and fine. Mouthparts well developed for chewing. Wings large, broad and with complex venation including many small crossveins. Cerci present or absent. Compound eyes large, sometimes taking up most of the head surface. Males with special structures on second and third segments of abdomen in which the sperm are deposited and from which they are transferred to the female during mating, which takes place in flight. Nymphs aquatic. Mouthparts of chewing type with labium developed into conspicuous organ (mask) for capturing prey, and which is folded below head when not in use. Gills of nymph internal within rectum or in form of expanded tail filaments. Powerful fast-flying insects (dragonflies) or delicate slower flying insects (damselflies).

Usually found near water, sometimes a long way from it. Nymphs in streams, ponds, lakes etc.

Division NEOPTERA

Metamorphosis hemimetabolous (Blattoid orders and Hemipteroid orders) or holometabolous (Endopterygote orders). Wings can be flexed over body.

BLATTOID ORDERS

7. Order PLECOPTERA
Stoneflies.

Antennae long, fine. Mouthparts weakly developed, adapted for chewing, sometimes reduced. Membranous wings held flat over back when not in use; hind wings larger than fore wings; venation complex or somewhat simplified. Tarsi of three segments. Cerci long, of many segments. Nymphs aquatic, with long antennae and cerci, gills usually present.

Found on stones and vegetation near water, seldom far from it.

8. Order GRYLLOBLATTODEA
Grylloblattids.

Wingless. Compound eyes small or absent. No ocelli. Antennae long and fine. Mouthparts adapted for chewing. Tarsi of five segments. Cerci long, of eight segments. Hemimetabolous metamorphosis.

Found beneath stones. Known only from the mountains in western North America, from Japan and Russia.

9. Order ORTHOPTERA
Locusts, crickets, katydids, grasshoppers, tree crickets, grouse locusts, pygmy mole crickets, king crickets, wetas and cave crickets.

Wings present or absent in adults; adults often with short wings. Mouthparts adapted for chewing. Prothorax usually well-developed. Tarsi usually with three or four segments. Fore wings (tegmina) narrower than hind wings and somewhat thickened. In immature stages the externally developing wings are reversed in position so that the future hind wing overlies the future fore wing. Females with large, conspicuous ovipositor. Cerci usually short, unsegmented. Sound-producing organs and special organs for hearing frequently present.

Found in many situations, in herbage, in or on the ground, under stones, in logs, in caves, and in sandy areas near water.

10. Order PHASMIDA
Stick insects, leaf insects, phasmids.

With or without wings. Mouthparts adapted for chewing. Prothorax short, but mesothorax and metathorax usually long. Tarsi of five segments. Fore wings usually small. In the nymphal stages the developing wings are not reversed in position. Ovipositor small. Cerci short. Sound producing organs and auditory organs not present.

Found on vegetation, to which their form and colouration usually bears a remarkable resemblance.

11. Order DERMAPTERA
Earwigs.

Antennae long. Mouthparts adapted for chewing. With or without wings. Fore wings in the form of small, short tegmina without any indication of venation. Membranous hind wings, broad semicircular with veins arranged radially. The hind wings fold away below tegmina in complex manner when not in use. Tarsi of three segments. Cerci modified into pincers.

Usually active at night. Found in soil, under bark, under stones, in foliage and flowers and in leaf litter.

12. Order EMBIOPTERA
Web-spinners, embiopterons.

Antennae long. Mouthparts formed for chewing. Fore and hind wings similar to one another; venation reduced. Males with wings, females without. Tarsi of three segments, the first segment of the prothoracic legs is enlarged, containing glands which produce silk. Cerci of two segments, asymmetrical in males. Hemimetabolous metamorphosis.

Found living, many together, in silken tunnels made by themselves. The tunnels are found under stones, in leaf litter, on rocks and on and under bark.

13. Order BLATTODEA
Cockroaches.

Antennae long, fine and of many segments. Mouthparts adapted for chewing. Legs adapted for running. Tarsi of five segments. Fore wings somewhat thickened, folded flat over body; sometimes wings absent. Cerci of many segments. No special organs for sound production or hearing. Eggs are laid in groups contained in special protective case (ootheca).

Found in houses, in logs, under bark, in caves and in leaf litter.

14. Order MANTODEA
Mantids, praying mantis.

Predatory insects with fine antennae, mouthparts adapted for chewing. Front legs adapted for seizing prey and the prothorax greatly elongated. Fore wings thickened and much narrower than hind wings. Eggs laid in a frothy mass which hardens to form an ootheca.

Found on vegetation, tree trunks.

15. Order ISOPTERA
Termites, white ants.

Social, polymorphic insects, living in large, subterranean nests which may be built into large structures above ground level. Each species has several forms, winged reproductive males and females and wingless, non-reproductive workers and soldiers of both sexes. Mouthparts adapted for chewing. Membranous fore and hind wings similar to one another; deliberately shed after nuptial flight. Tarsi usually of four segments. Cerci short.

Found in ground nests, very seldom in the open. Conspicuous nuptial flights take place at certain seasons, usually towards dusk.

16. Order ZORAPTERA
Zorapterons.

Winged or wingless individuals of both sexes known in the same species; not truly social although several may be found together. Antennae of nine segments. Wings can be deliberately shed; venation much reduced. Prothorax large. Tarsi of two segments. Cerci unsegmented, short.

Found in dead wood, decaying leaf litter, under bark or in termites' nests.

HEMIPTEROID ORDERS

17. Order PSOCOPTERA
Psocids, book-lice.

Antennae long, fine, with thirteen to more than fifty segments. Mouthparts adapted for chewing but with the lacinia modified to form a unique chisel-like structure. Tarsi of two or three segments. Wings membranous, fore wings larger than hind wings. Venation reduced; wings held roof-wise over abdomen when not in use.

Found in houses, on or under bark, under stones, in leaf litter, on leaves and in stored products, sometimes in compact groups or under silken webs; some domestic species are pests in collections of dried insects.

18. Order MALLOPHAGA
Biting lice, bird lice.

Small, wingless flattened insects parasitic on warm-blooded animals, mainly birds. Antennae short. Compound eyes small. No ocelli. Mouthparts modified from the chewing type. Prothorax fairly large. Tarsi unsegmented or of two segments. No cerci.

Found mainly on birds.

19. Order SIPHUNCULATA
Sucking lice, lice.

Small, wingless, parasites of mammals. Antennae short, of three to five segments. Compound eyes small or absent. No ocelli. The mouthparts, adapted for piercing and sucking, can be withdrawn into the head capsule. Segments of thorax fused. Tarsi unsegmented. No cerci.

Found on mammals.

20. Order HEMIPTERA
Plant bugs. Cicadas, frog-hoppers, cuckoo-spit insects, tree hoppers, leaf-hoppers, lanternflies, psyllids, lerps, whiteflies, aphids, greenflies, plant lice, scale insects, lac insects, mealy bugs, ground pearls (these are included in suborder Homoptera), assassin bugs, pond skaters, water striders, capsids, bed bugs, cotton stainers, lace bugs, toad bugs, shield bugs, water bugs, water scorpions, back swimmers and water boatmen (these are included in suborder Heteroptera).

A large and variable order. Mouthparts modified for piercing and sucking; the palps are reduced and the labium forms a trough in which lie the stylet-shaped mandibles and maxillae. Four wings usually present, wingless and polymorphic species known. Fore wings somewhat thickened throughout (in suborder Homoptera) or only in basal section with the distal part remaining membranous (suborder Heteroptera). Venation very variable. Compound eyes and ocelli usually present.

Found mainly on vegetation; but members of this order may be found in almost any situation; many species are aquatic.

21. Order THYSANOPTERA
Thrips.

Small. Antennae short and of six to ten segments. Mouthparts modified for piercing and sucking, asymmetrical. Tarsi unsegmented or of two segments, with a protrusible sac at the distal end. Wings narrow, almost strap-like, with greatly reduced venation and long marginal fringe of hairs. No cerci. The change from nymph to adult is via one or two inactive instars which may be regarded as incipient pupal stages.

Found on fresh vegetation, sometimes in dried leaves and flower heads.

ENDOPTERYGOTE ORDERS

Metamorphosis holometabolous. Pupal stage present. Developing wings internal, not visible externally in young stages (larvae). Larval and adult body forms differing greatly from one another; larvae and adults usually with very different habits.

22. Order NEUROPTERA

Spongillaflies, alderflies, brown and green lacewings, antlions and snakeflies.

Small to large insects. Antennae long. Mouthparts adapted for chewing. Fore and hind wings similar to one another, membranous, held roof-wise over abdomen when not in use. Venation fairly complete with many additional veins. No cerci. Larvae carnivorous, with mouthparts adapted for chewing or for sucking body juices of prey.

Found in a variety of situations, usually on vegetation; some families have aquatic larvae and the adults of these are found near water.

23. Order MECOPTERA

Scorpionflies.

Small to moderately large, long-legged insects. Antennae long and fine. Head usually extended downwards into a rostrum. Mouthparts of the chewing type. Fore and hind wings similar to one another, held over the back when not in use. Venation fairly complete. Cerci short.

Found in a variety of situations, amongst herbage.

24. Order LEPIDOPTERA

Butterflies and moths.

Antennae long of various forms, often differing in the two sexes. Wings membranous. Venation specialised with few crossveins linking main veins. Body, legs and wings clothed in scales. Mandibles usually absent and the maxillae forming a tube used as a sucking proboscis. Pupae usually within a silken cocoon spun by the larva before final larval moult or within an earthen cell.

Found almost everywhere.

25. Order TRICHOPTERA

Caddisflies.

Antennae very long and fine, tapering. Mandibles reduced or absent. Wings are membranous, held roof-wise over abdomen when at rest and thickly covered in hair. Fore wings longer and more slender than the hind wings. Fairly complete venation with few crossveins linking main veins. Tarsi of five segments. Larva aquatic, caterpillar-like, usually bearing a case of its own construction into which it retires when disturbed.

Found in the vicinity of water in which the larvae live, usually on stones or waterside vegetation.

26. Order DIPTERA

Two-winged or true flies, daddy-long-legs, crane flies, sand flies, moth flies, mosquitoes, midges, gnats, gall midges, fungus gnats, bee flies, hover flies, vinegar flies, fruit flies, warble flies, bot flies, blow flies, blue bottles, green bottles, cluster flies, stable flies, house flies and tsetse flies.

A large, variable order. Only one pair of functional wings, the hind pair being modified to form the halteres. Antennae may be long or short. Mouthparts adapted for sucking, and in some families also for piercing. Tarsi usually of five segments. Wing venation variable, usually with few crossveins. Larvae caterpillar-like or maggot. Some flies parasitic. Found everywhere.

27. Order SIPHONAPTERA

Fleas.

Antennae short, held in grooves in the head capsule. Small jumping insects, ectoparasitic on warm-blooded animals. Laterally compressed. No compound eyes; usually two ocelli. Mouthparts strongly modified for piercing and sucking. Coxae enlarged. Tarsi of five segments. Wings absent.

Adults found on warm-blooded animals; larvae in vegetable and animal debris.

28. Order HYMENOPTERA

Wood wasps, horntails, sawflies, ichneumonflies, braconids, gall wasps, ensignflies, fig insects, fairyflies, velvet ants, ants, wasps, mud daubers, potter wasps, social and solitary bees.

A large order of insects with four membranous wings, with reduced venation. Hind wings smaller than fore wings and attached to the fore wings in flight by means of a row of hooks (hamuli). Antennae usually long. Mouthparts adapted for chewing and lapping. First segment of abdomen fused to metathorax; abdomen narrow near base to form a distinct 'waist'. Ovipositor modified to form a sting, or used for piercing or sawing. Larvae usually grub-like, sometimes caterpillar-like.

Found everywhere. Some polymorphic species live in colonies; many species parasitic in their larval stages within the bodies of other insects.

29. Order COLEOPTERA

Beetles.

The largest order in the kingdom. Of great variety in size and form with fore wings modified into hardened elytra under which the membranous hind wings are folded when not in use. Venation virtually indistinguishable in elytra, hind wings with reduced venation or absent. Mouthparts adapted for biting. Antennae long. Prothorax large, mesothorax reduced in correlation with reduction of flight muscles. Larvae of great variety of form including all types.

Found in almost every situation.

30. Order STREPSIPTERA

Stylops.

A small order of species parasitic in their larval stage on other insects. Adult males winged and free-living; adult females degenerate, remaining in host insect within a puparium. Antennae conspicuous. Mouthparts degenerate, derived from a type adapted for chewing. Fore wings reduced to small club-shaped organs, hind wings large. Larvae undergo hypermetamorphosis, i.e. several changes of form occur in the course of development.

Adult males are occasionally attracted to light. Females are found in their hosts (usually wasps and in some families of sucking bugs) protruding from the body of the host.

CHAPTER 3

COLLECTING AND TRANSPORTING INSECTS

THE EXTREME ADAPTABILITY WHICH the basic plan of insect anatomy allows has led to the insects evolving into one of the most successful groups of animals and insects are found in almost every part of the world. Only in the sea do they not abound but, even there, there are to be found a few specialised species which have managed to overcome the problems of living in a salt water environment. Quite apart from their ability, as a group, to cope with extremes of conditions insects have become adapted so as to thrive on a tremendous variety of foods. There are few species of plant which are immune to insect attack; creatures of almost every class which can live on land or in fresh water are attacked by them and the insects themselves, by virtue of their abundance, make an excellent source of food for other, predacious, insects.

They are more abundant in spring, summer and early autumn than they are in winter. In the tropical areas in which there are pronounced dry and wet seasons they tend to be most abundant when wet and warm weather coincide. Even in the coldest weather they can be found by searching in the right places. There are species which are active only during the day; there are those which are active only at night and some restrict themselves to being active at dusk and dawn. There is always something for an entomologist to find and collect no matter where he is. Insects may be found flying free in the air, on foliage, in and on flowers, in fruit, within the living wood of plants, on or under bark, at the roots of plants, in the leaf litter which lies on the forest floor, in the soil, under stones, in the droppings and carcasses of animals, in the water and bottom mud of streams, ponds and lakes and they may be found in or on the bodies of other animals, including the bodies of other insects.

In many cases the best way of finding insects is to simply search for them individually. They can be seized with the fingers or by means of a pair of fine-pointed forceps. By searching for individual insects you can often come across an insect going about its business undisturbed and so come to learn a little of its habits. This method will yield relatively few specimens and the use of some piece of collecting equipment will increase the numbers taken and facilitate their handling.

The equipment need not be elaborate. For some specialised work collectors have designed elaborate and sometimes costly pieces of equipment, but a very good collection can be made with the simplest of equipment. Much of this can be made at home and the remaining items are not very expensive.

BASIC EQUIPMENT

The basic equipment which is usually carried on a general collecting excursion is listed below. There are no hard and fast rules about what you will need and with experience you will become interested in one particular group of insects or in one particular aspect of insect life. You will then require only equipment which is suited to your interest. A list of basic requirements would run something like this:

Collecting net
Aspirator
Forceps
Camel-hair brush
Killing bottles
Pill boxes
Glass vials of various sizes with stoppers, some dry, some containing liquid
    preservatives
Miscellaneous boxes and tins
Box with layers of tissue paper
Paper triangles or small envelopes
Paper labels
Note book
Pencil
Magnifying glass
Collecting bag in which to carry above items of equipment.

Later in this chapter we shall discuss the techniques employed in collecting in special habitats and some of the other pieces of equipment which will then be useful will be described.

Collecting net

The traditional insect collecting net is still the basic piece of entomological equipment, although there are a great many nets differing from one another in small points of design. Essentially, a net consists of a wooden handle, with a metal ring at one end from which is suspended a cloth bag.

Handles may be made of wood, such as a broom handle, cane or aluminium tube. A tube has the advantage of greater strength for equivalent weight and the lighter the net the better. The length varies according to the uses to which the net is to be put; a handle of between 90–120 cm in length is convenient for most purposes; handle diameter should be such that it is

The traditional insect collecting net.

comfortable to hold for long periods. A diameter of a little less than 2.5 cm is generally found to be quite comfortable. By the use of metal ferrules it is possible to make sections of handle which will fit into one another end to end and to obtain handles of any length. A handle of more than about 400 cm is unwieldy. Individual sections should not be much longer than 120 cm.

The material of which the ring is made also varies. Steel wire about 5 mm thick is excellent for general purposes. Lighter wire is more easily bent but is adequate if the net is not to be used for heavy duty. Flat, flexible steel band may also be used. It is also important to arrange the ring so that it can be collapsed, this will make it easier to carry when not in use. A convenient form of net can be purchased from fishermen's supply firms, usually called a landing net. The circle may be made of four arcs each hinged to the next by means of a rivet. The ring can then be collapsed to form quite a small parcel with the cloth bag still in position. This method of folding is best used on the heavier gauge rings as a stronger hinge rivet can be used. In the usual collecting net the ring is anything from 30 cm to 45 cm in diameter but many collectors use larger diameters. A simple way of attaching a wire ring to a wooden handle is as follows. The open ends of the ring are bent at right angles to the circle, outwards, and again at right angles inwards towards each other a few centimetres further along, a short distance from the ends of the wire. A groove is cut in each side of the handle, at the ring end, with a short hole at the end of each groove away from the end of the handle. These grooves and holes will accommodate the wire ends of the ring; the wire should fit neatly into the grooves. To keep them in place a metal ferrule is slipped up the handle over the wires. This ferrule, also, should be tight-fitting so as to hold the wires firmly into the grooves. Another method of attaching a tubular ring to a metal handle is illustrated.

A fine mesh, lightweight material should be used for the net itself. If the cloth is tight-woven the net will not sweep easily through the air; it is also an advantage to be able to see through the net from the outside so green or black material is better than white. Bretonne net, marquisette or cotton netting is

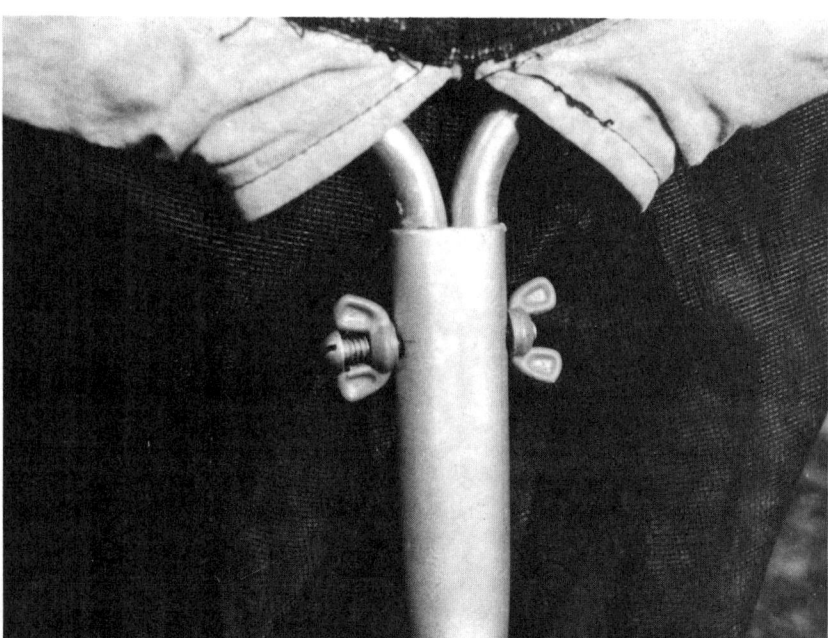

One method of attaching a net to a tubular metal handle.

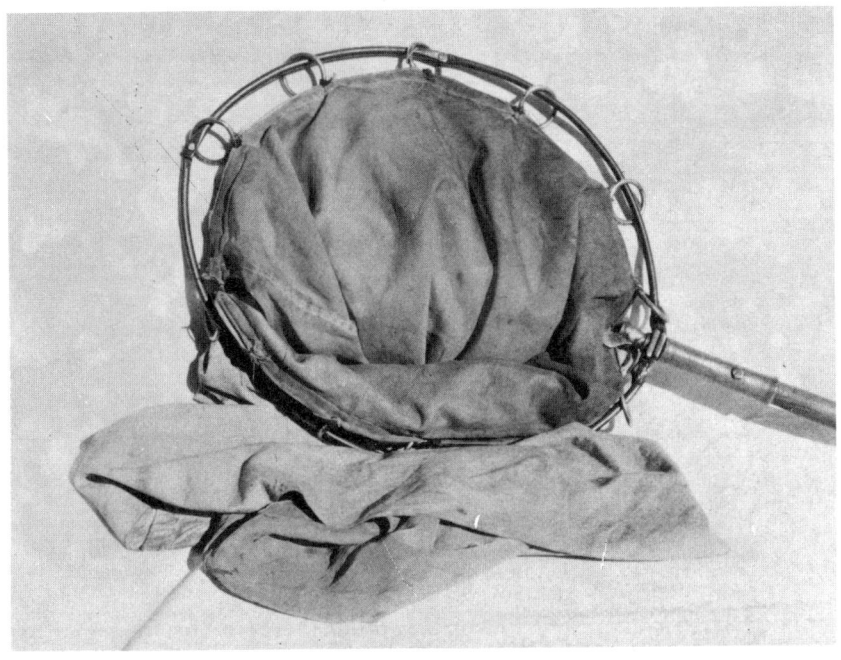

A sweep net, a fine net protected by a heavy cloth bag.

useful; a variety of synthetic fibre netting is available. For a sweep net, which has to take rough treatment, unbleached calico, lawn or a closer-woven synthetic fabric will be found most durable or a calico sleeve can be slung from the ring to protect a more delicate net.

The net itself should be tapered to a rounded end at the bottom. Its length should be about twice the diameter of the metal ring and the circumference of the mouth of the bag should be made a little larger than the circumference of the ring which will support it; this will give it a loose fit on the ring. The material of the bag should not be sewn directly onto the ring. A light canvas or thick calico binding, a few centimetres in width, should be sewn to the top of the net and the ends left open so that the metal ring can be passed through the binding. In this way the net can be removed and replaced.

Use of nets

An ordinary net is used to collect individual specimens when they are in flight or at rest on flowers or herbage. A simple, fast, accurate sweep of the net will usually obtain the specimen. It is quite often possible to net specimens by sweeping the net from behind them as they fly. Some species seem singularly adept at avoiding a net when they fly towards it. If the material of the net is too thick it is possible for a pocket of air to be retained in the net which deflects the specimen. When the specimen is at rest it is wise to bring the net as closely as possible to it and then make a sudden, rapid sweep. A sideways and slightly upward sweep is preferable to a down-stroke. A specimen on the ground can be caught by simply dropping the net over it. It will usually fly up into the net. Once the specimen is in the lower part of the net, the net handle should be quickly twisted; the lower end of the net will then fall over the rim of the ring. This will usually trap the specimen in the folds of the lower half of the net from where it can be taken into a pill box or other container or directly into the killing jar. It is preferable to insert the receptacle into the net and manoeuvre it within the folds of material until the specimen is safely caught. The box or bottle should be closed whilst within the net. When a net

The position of the net at the end of a stroke, the insect will be trapped in the folds of the net.

of about 40 cm in diameter is laid on the ground there is ample room to work with both hands inside.

A sweep net is used in a different way from an ordinary net. By means of a sweep net a sample of the insects on a tree, shrub, or in grass can be obtained. A sweep net is used with a sideways sweeping motion, passing the rim of the net through the vegetation with bold strokes. The rim of the net jars and brushes through the vegetation thereby disturbing and dislodging the insects. They fall or fly into the mouth of the net and accumulate on the sides and in the bottom of the net. The sweep net is particularly useful for collecting small species, of which there is a multitude, which would otherwise never be seen. The net should be inspected after every few sweeps. Plant material will become dislodged and fall into the net and if sweeping is continued for too long without removing the catch and discarding the debris the specimens will become crushed.

The contents of the net should be removed by hand, forceps or aspirator or the insects caught can be taken directly into glass vials. If the contents are shaken to the bottom of the net and the net held closed just above them, a small narrow passage way through the folds will enable many of the insects to creep up and out of the net, when they can be taken. Careful opening of the folds will reveal other specimens which can be persuaded into bottles or taken up by an aspirator. The debris which is left behind can be searched for any remaining specimens. Another way of dealing with the sweep-net catch is to shake the contents of the net to the bottom and invert this into a killing bottle, debris and all; this is sorted out later. This method has the disadvantage that the debris makes the killing jar dirty and also introduces excess moisture from the plant material into the killing jar. It also kills some insects which otherwise may have been allowed to escape. If preferred, the end of the net, with its contents, can be placed in a large killing bottle without inversion and the lid of the jar held in position until the insects in the net are dead. This method has the disadvantage that it takes a little time for the poisonous fumes to penetrate the bag and take effect. The contents of the net are emptied out onto a sheet of paper and sorted. Care must be taken in sorting through debris not to miss small specimens.

Sweeping wet foliage usually yields few specimens and the net, once wet, becomes heavy and difficult to manage. Specimens stick to a wet net and are usually ruined.

The aspirator bottle

The aspirator bottle is a device for collecting small insects individually. It can be used for taking insects directly from the ground or from foliage etc. or from the net or beating tray (described later).

An aspirator bottle consists of a bottle (7 cm by 2.5 cm is a good size) fitted with a cork, or preferably, a rubber stopper. Through the stopper are bored two holes to take two pieces of glass tubing, each about 7 cm long and about 5 mm internal diameter. Each piece of tubing is bent in the middle. One piece of tubing is pushed through each hole in the stopper, almost as far as the bend in the middle; wetting the tube will make this easier. The end of one of the pieces, which will be inside the bottle when the stopper is inserted into it, should be covered over with a piece of cloth, such as organdie or muslin and should be tied securely. The other tube is left open at both ends. To the opposite end of the piece of glass tube to which the muslin is tied, attach a 30 to 45 cm rubber or plastic tube. Replace the stopper, with its glass tubes, making sure that the stopper is tight-fitting and airtight and that the glass tubes fit tightly in their passage through the stopper.

To use an aspirator place the open glass tube close to the small insect to be captured. Put the end of the rubber tube in your mouth and suck sharply; an indraught of air will suck the insect up and into the bottle. The muslin over the end of the second tube in the bottle prevents the insect reaching your mouth. With a little practice this instrument can be used for collecting large

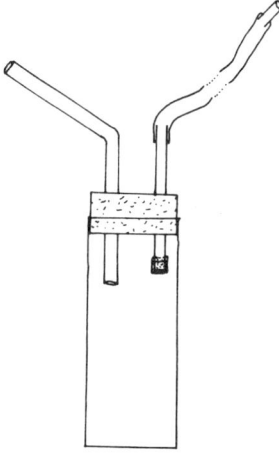

An aspirator (sometimes called a 'pooter').

The aspirator in use— collecting small insects from a beating tray.

41

numbers of small insects. They can be shaken from the aspirator into the killing jar.

Forceps

For collecting, the best type of forceps are those of which the prongs are rounded or come to a point with the internal surfaces milled. The best length is about 10–13 cm. When purchasing forceps it is a good idea to buy two or three pairs as they have a habit of getting lost. A length of brightly coloured ribbon attached to them will assist in finding them. The prongs should come together *at the tips* and not before so that when an insect is taken between the tips it will be gripped and not allowed to slip out.

Camel-hair brushes

Two camel-hair brushes, of different sizes are useful for collecting small insects, especially those which are to be preserved in liquid preservative. Size 2 and Size 5 are commonly used. If the tip of the brush is moistened in the preservative and dabbed on the insect it will adhere to the brush. If the brush is again put into the preservative the insect will float free of the brush. With a little practice a collector can become adept at picking up quite small specimens without any debris.

Killing bottles

It is convenient to have several killing bottles on hand when collecting. One should be large, and the others smaller. The most efficient killing agent is potassium cyanide but the most generally used is ethyl acetate. As cyanide is a deadly poison it is not always easy to obtain. A chemist may make up a killing jar when the purpose for which it is to be used has been explained. Laws vary regarding the sale of cyanide and a signature may be required before a purchase is made. A large killing jar can be made from any wide mouth glass jar with an *airtight* stopper or screw-on lid. About 20 mm of crushed cyanide is placed in the bottom of the jar and covered by a shallow layer of sawdust. A sheet of paper, cut to fit the shape of the jar (which should be round for preference), is placed on top of the sawdust and on top of this is poured a mixture of plaster of paris and water to a depth of about 10 mm. The plaster of paris should be made by adding powdered plaster of paris to water, little by little, stirring all the while, until the mixture has the consistency of thick cream. It should then be poured onto the middle of the layer of paper in the bottle and allowed to spread out to reach the sides of the bottle. Avoid spashing the sides of the bottle and try to avoid bubble formation. The bottle should be dried out completely in the *open air*, in the shade. Crumple up some tissue paper and almost fill the bottle loosely with this, then close the bottle. It is now ready for use. A series of smaller bottles can be made in the same way. When making a killing jar great care should be taken not to spill cyanide and the hands should be washed thoroughly with soap and water afterwards, special attention being given to cleaning the finger nails.

Cyanide killing bottles should be clearly marked 'Poison' and should *never* be left in a place where children can find them. They should always be under lock and key when not in use. Stocks of cyanide should not be kept, but a fresh supply obtained when bottles are to be made.

Another type of bottle is the ethyl acetate bottle. A layer of about 25–40 mm of plaster of paris is poured into an empty jar and allowed to solidify and dry thoroughly. When this is completely dry it is saturated with ethyl acetate. Ethyl acetate is highly volatile and tight corking is essential.

The bottle is now ready for use. Ethyl acetate is more readily obtainable than cyanide.

Instead of using ethyl acetate, carbon tetrachloride, benzol or ammonia can be used in the same type of bottle.

Other popular killing agents are chloroform and ether. For these it is not usual to prepare a special bottle; a wad of cotton wool with a few drops of ether or chloroform are merely dropped into the bottle containing the specimens to be killed and which is then tightly corked.

The above killing agents are usually used for specimens which are to be preserved dry. Many insects, especially larvae, are preserved in liquid. Killing these is dealt with later; at end of Chapter 4 will be found details of the methods suggested for each order.

Using the bottles

All the killing agents kill by means of poisonous fumes and it is therefore essential that all killing jars be kept tightly stoppered at all times except when insects are actually being put into the jar. This will minimise wastage of poison and chances of accidentally inhaling the poisonous fumes. Killing jars should be kept out of the sun whenever possible.

Killing jars should be wiped out to prevent condensation of moisture and remove dust. Make sure that the fumes kill in as short a time as possible. A cyanide bottle that is weak should be destroyed by breaking and burying. An acetate or carbon tetrachloride bottle can be dried out and recharged.

In order to get an insect from the net into the jar insert the open jar into the net. Force the insect into the bottle and close the bottle with hand or thumb; the insect should fall down almost immediately; the bottle can then be withdrawn from the net and corked. If the insect is large and the knock-down effect is not immediate, place the bottle against the net from the inside and apply the cork from the outside. When the insect is sufficiently quietened take the bottle out of the net and cork it. Use this method if the insect is likely to sting or bite. Do not leave insects in the killing bottle for longer than necessary; some of the larger insects such as large beetles or grasshoppers may need some time to die although they may become immobile soon after being put into the jar. It is advisable to put moths and butterflies into a special bottle of their own as they give off large quantities of scales from their bodies and wings which adhere to other insects as well as making the bottles dirty. It is also advisable to empty the killing jar at frequent intervals as the greater the accumulation of insects and the longer they are left in the bottle the more moisture is given off.

Ethyl acetate does not kill as quickly as cyanide but usually leaves the insect more relaxed; colours are not usually affected. Ether and chloroform must be handled with care and although they have a quick knock-down effect may take a little longer to kill; they are both extremely volatile. Insects vary in their degree of stiffness after killing by these two agents. Carbon tetrachloride is a slower killer but shares with ethyl acetate the quality of being relatively harmless to human beings; it is not as volatile as ethyl acetate and, while its effects on insects after death vary, colours do not usually appear to be affected to any great extent.

It is probably true to say that cyanide is the best all-round agent, providing the dangers attending its use are realised; ethyl acetate is next in line with the advantage that it is less toxic to humans.

Pill boxes

Pill boxes, of the old fashioned type once obtainable from chemists, and now obtainable from dealers in natural history supplies, are very useful for collecting individual specimens, such as butterflies, from the net. As they are wide-mouthed, coming in a wide range of sizes, specimens can be guided into them quite easily and the lid slipped on. They are light-tight, and butterflies and other insects will usually remain quite still inside the box after capture so tending not to damage themselves. The specimens can be kept in the pill boxes and be killed just prior to pinning and setting, processes which are described in Chapter 4.

Glass vials

Glass (or plastic) vials are useful for holding individual large specimens or several small ones after capture. The collecting kit should include several, preferably in a range of sizes, some of which should be clean and dry while others should contain liquid preservatives. Liquid killing and preserving agents are discussed in the next chapter. They are used for killing and preserving many of the soft-bodied insects, especially small species, and larvae. The dry bottles are used for carrying home alive individual specimens for rearing or later study. The vials should be stoppered; many small insects will survive for quite a while in a stoppered bottle provided that they are not overcrowded and provided also that moisture from their bodies does not condense on the sides of the glass. By putting some loosely crumpled tissue paper in the vial the risk of this will be lessened. If larger specimens are to be transported alive, a hole can be made in the cork or stopper and this covered over with fine gauze to allow passage of air. These can be bought from entomological supply houses. Should the insect be required for rearing, a little of the appropriate food should be placed in the vial as well. Glass vials should not be left in the sun as this will soon kill the insects.

Glass vials can also be used for bringing home small dead specimens after removal from the killing jar. A small wad of tightly packed tissue paper is placed in the bottom of the vial over a few crystals of chlorocresol. The dead insects are placed on top of the wad with a data label (see below) and another wad of paper is placed on top of them and pressed down to prevent the specimens from moving but not hard enough to squash them. In this way layers of insects and paper wads can be built up until the vial is filled. Plastic vials tend to be scratched easily and eventually crack. Also, many plastics are affected by ethyl acetate or chloroform.

Miscellaneous boxes and tins

It will be found quite useful to have a small supply of miscellaneous tins and boxes, such as cigarette tins or cough lozenge tins, on hand. These are used in the same way as pill boxes and glass vials and are especially useful for carrying home live material, larvae, pupae etc.

Box with tissue paper layers

One of the important items of field equipment is a box or tin, preferably of cigar box size or a little smaller, containing layers of soft tissue paper cut to size so as to fit snugly into the box and fitting well into the corners. When the first lot of insects is ready to come from the killing jar most of the layers of tissue paper are removed leaving a few layers in the bottom. The dead insects are emptied onto this and a data label placed with them, the rest of the tissue paper layers are then put on top. When the second batch of specimens is ready the same procedure is carried out, but each batch of insects is separated from the previous one by a few layers of tissue paper. In this way the box is filled with alternating layers of tissue paper and insects. If there is a wide range of sizes in each catch, put the larger specimens in a layer of their own separated from the smaller specimens. Each layer must receive its own data label and specimens from separate localities must not be mixed. It is important to make sure that the insects are held firmly in place between the tissue paper layers by the pressure of the layers above, but they should not be crushed.

If the insects are to be dealt with at home the same day no further action is required. If not, some crystals of chlorocresol should be placed in the bottom of the box, below the layers of tissue paper and when the box is filled it should be sealed so as to be airtight. Under these conditions the insects will remain fairly supple due to the fact that they do not dry out.

It is *not* a good idea to use cotton wool or any other fibrous material in place of soft tissue paper layers as the insects are caught up in the fibres and are difficult to extricate.

Paper triangles and small envelopes

A few envelopes, preferably of cellophane, are handy. Some insects are better stored in these, especially those which shed their limbs easily, such as the craneflies. These often lose their limbs in the killing jar and are best put into a small cellophane envelope with their loose limbs and a data label after they are removed from the killing jar. Paper triangles are used for butterflies and moths which lose the scales of their bodies and wings. If placed in tissue paper in a box there is a likelihood that slight movement will rub them and remove the scales. As their colour patterns are derived from the arrangement of the scales this is most undesirable. Instead, each specimen, after killing, is placed in a paper triangle which is made from a rectangular piece of paper. The wings and legs of the specimen can be arranged so as not to be distorted when the specimen is placed in the triangle. Its subsequent handling will then be quite easy. The triangles can have the relevant data (see below) written on them and they can be packed flat in a shallow box, to transport home.

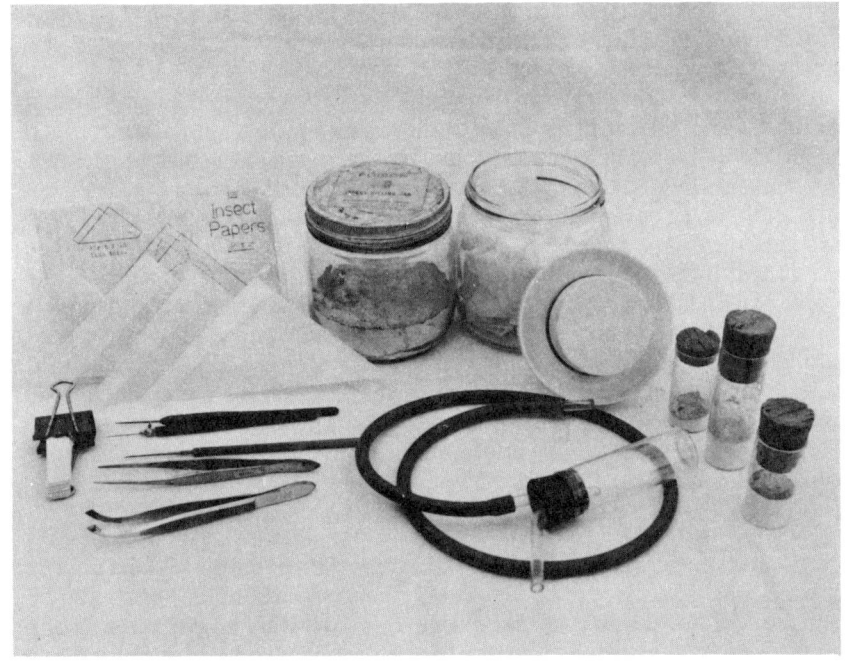

Some of the basic pieces of equipment used in insect collecting.

Paper data labels

When finally placed in the collection each specimen, or, in the case of specimens preserved in liquid, each bottle, must be labelled with information relating to its capture. This information must be recorded at the time of capture and it is convenient to have some small paper labels, ready-cut. A convenient size is about 50 by 15 mm. This size is suitable for putting into bottles with liquid preservative, with each layer of insects between tissue paper and into paper envelopes. If these labels are ready-cut they can be held in a small spring clip such as used in offices for holding papers together. It is advisable to use a fairly good quality paper, even for field labels, as the information written on them is of great importance.

The basic information required for each and every specimen taken includes the following: place of capture, date of capture and name of collector. The more precisely the locality of capture is defined the better; in many cases it is not possible to be precise because in the country the collector may not know exactly where he is and may have to use some such indication as '40 km from such and such, on road to so and so'; a map grid-reference may be useful here. It is also useful, if possible, to give an indication of the altitude at which the specimens were taken. The locality should be indicated so that someone not familiar with the country concerned can locate at least the area on a map, the county or state should be given as well as town locality. References to suburbs of cities should not be made without references to the city. The date is usually written thus: 12 Jan 1963, or 12.I.1963 or 12.i.1963. Some American collectors reverse the day and month; the above example would then read 1.12.1963. This can be confusing and as the former method has been used for many years on a world-wide basis it is strongly recommended that it be retained in preference to the newer, less widely-used American system. The name of the collector should include initials. A specimen without adequate information may be a pretty object but it has no scientific value whatsoever and may be a source of confusion. *The label information must be recorded in the field* at, or soon after the time of capture; never trust to your memory when it comes to label data. Specimens from different localities and dates should always be kept separate from one another and labelled separately.

It is very often useful to have additional information. The most usual additional information recorded is the host plant or host animal from which the insect was taken; in the case of parasites the name of the host is essential. Also, precise indications of habitat are useful, such as 'under stone', 'on bark' etc. Never be afraid of putting down too much information.

If specimens are baited or trapped, the bait or type of trap should be mentioned. If the specimens are reared, the fact should be mentioned. Some people place a label with a serial number only with the specimens and make a full note of the necessary data in a field notebook opposite the same serial number. This method should not be used. If a collecting trip of any length is undertaken and the book is lost before permanent labels are prepared all the specimens collected will be without data and virtually worthless. It is, of course, quite in order to put additional information which cannot be put onto the label into such a serial book.

Notebook

It is useful to have a general notebook to use as a field diary. Make a note of the date, time and place of each excursion; even short excursions should be included, not only major collecting trips. It is surprising how frequently one wants to know odd details about a trip undertaken some time before, but memory is notoriously unreliable in supplying details. Worse, even, is that you sometimes feel quite sure of details but find on looking them up that you were quite wrong. If you are accompanied on a trip by someone else, make a note of it, too. A word or two about weather conditions can be put in. The field notebook should preferably be of pocket size, about 15 by 10 cm and should be hard-covered. The cover should be dyed with a fast colour that will not run if the book gets wet. The pages should be lined and if you number the pages in sequence it is easy to index the notebook. When each book is full it can be given a volume number and one index can be maintained to cover all the volumes. This can be kept up to date as you go along by giving reference to volume and page number. For example, you may refer to a certain species on several occasions in various volumes, in which case that species can be mentioned in your index followed by the various volume and page numbers on which reference to it can be found. As the years progress, snippets of information collected in the field in quite casual fashion may turn out to be of considerable value. Some entomologists keep rough field notes which are written up into a fuller field diary at home; the volumes of this fuller diary are kept. This method has the advantage that information can be better organised in the final written-up diary than in the field notes.

Pencil
Writing in the field is important and it is best to have a good quality B or HB pencil for this. Ball-point type pens are usually not satisfactory.

Magnifying glass
A magnifying glass is an essential piece of equipment. They come in various qualities and when choosing an instrument two factors are to be considered: one is the actual magnifying power of the lens and the other is the width of the field of vision. Generally, the greater the magnification the smaller the field of vision, unless this is compensated for, in which case the cost of the instrument rises sharply. A good quality $10 \times$ or $12 \times$ is satisfactory for most purposes. It is worthwhile buying a good instrument in the first instance.

Collecting bag
Collecting is not always done on flat or easy terrain. It is necessary to have your hands free and carrying collecting equipment is a problem which can only be solved by keeping it in some sort of bag. It is useful to have a canvas bag, or one made of some other strong material, with partitions which will enable you to make a rough sorting-out of the items in it. If your collecting becomes specialised, as it probably shall, you may be able to arrange pockets or holders within the bag for each item and you will then be able to put your hand instantly on any required item. A rough sorting, however, is quite adequate when general collecting is being undertaken. Do not put foodstuffs to be eaten on an excursion in with killing jars. As it is important for your hands to be free, the bag should be slung from the shoulder by means of a strap. If the strap is put over the head and the right arm put through the sling the bag can be made to hang near the right hip. This is the most accessible position and the bag can be slung forward when needed and back when not in use. A bag slung on the back has the disadvantage that it must be taken off each time it is needed.

COLLECTING IN VARIOUS HABITATS
The success of insects in colonising the world has led to their occupying almost every habitat. In many cases, special techniques can be used which are

more efficient in time and yield of specimens than general techniques. The general collector may find some of the techniques described below useful as they will enable him to collect more efficiently in some habitats than his ordinary techniques will allow.

Collecting insects in flight

Only adult insects will be taken in flight. For this the butterfly net will usually suffice as most will be captured individually and will be relatively large. Flying insects are, of course, met everywhere and netting will normally be done during the day although it is often possible to obtain nocturnal insects in flight when they are attracted to lamps, especially mercury vapour lamps. An extension handle is useful for collecting out over water and also at street lamps as the insects sometimes fly as high or higher than the level of the lamp itself. For more efficient night collecting a light trap, described later, can be used. As the insects taken in flight are usually relatively large they can be put into the killing jar one by one as they are caught.

Collecting from foliage

Collecting from foliage is probably the most rewarding of all so far as numbers of specimens are concerned because vegetation harbours a great wealth of insect life. Searching for individual insects can be very interesting and is usually the only way in which insect eggs or egg masses will be found. If an insect is noticed it should be watched for a moment before being taken so that any interesting activities can be noted. In searching, it is as well to be systematic, running the eye along the leaves of one branch, then casting the eye back to the starting point and following out another branch. Bigger insects are collected by hand or forceps; smaller specimens are taken by aspirator or by sweep net.

The beating tray is a development of the 'inverted umbrella' which is still used by some entomologists. In this method of collecting an umbrella is held open, upside down, beneath the branches of trees and shrubs and the branches are beaten with a strong stick. This dislodges the insects which either fly off or fall into the upturned umbrella. The specimens are collected by hand, forceps or aspirator. A beating tray is used in the same way as an

Using a beating tray.

One form of beating tray—upper side.

Beating tray—lower side.

umbrella but consists of a framework of four radiating arms and two handles. The arms form the supports for a square piece of cloth and are made of malacca cane or 12 mm aluminium tube (both these materials combine strength with lightness). The handles are made of wood. The handles and arms are held together near one end by a nut and bolt upon which they all swivel. The cloth is square with a split from the middle of one side to the centre and is bound along either side of the split with a binding of about 5 cm in width. The wooden handles pass through these bindings and extend to a little beyond the edge of the cloth. The radiating supporting arms slide, at their free ends, into sleeves at the four corners of the cloth sheet. The handles are held together by a clip or ring to form one handle and in this position hold the cloth taut. The tray may be collapsed by parting the handles and swinging them through 180°. The whole thing then forms quite a convenient bundle for carrying. The best length for the handles and radiating arms is 75 cm. This frame will support a cloth tray of about 90 cm square. Some people prefer black cloth rather than white. Beating the vegetation causes many species to take flight, such as the flies, wasps and other active insects but the less active forms, such as beetles, psocids and many bugs, thrips etc. are best collected by beating tray. It is as well to pay attention to bunches of dried leaves hanging from otherwise green trees. Such dead leaves support a fauna which is very different from that which the same leaves would have when fresh. A simpler type of beating tray is also illustrated.

A simpler form of beating tray—upper side.

Simpler beating tray—lower side.

Leaf miners

Many insects live within the tissues of leaves, in leaf mines of their own making. These mines are usually visible as serpentine markings on the leaves or as blister-like blotches. The insects causing these mines are larvae of various orders of insects and in order to obtain their adults the leaves must be picked, preferably with some stem, and kept as fresh as possible in a container so that the insect can complete its development to the adult which will then emerge.

Collecting from flowers

As flowers develop from the bud stage and pass on to the fruit or seeding stage, they have associated with them a changing insect fauna. Insects may be found living in the buds; to collect these the buds must be placed in containers so that the adults of the developing insects can be taken when they emerge. Many small species are found only within flowers; many others visit the flowers to collect nectar, pollen or to find mates which are attracted to the flowers for feeding. A sweep net or butterfly net is useful for collecting the big, obvious visitors to the flowers. To obtain the small habitual flower inhabitants the individual flowers or groups of flowers should be held out over a sheet of white paper or cardboard, or over a white enamel tray or beating tray and shaken. The insects which fall out can be collected by aspirator. In this way less damage is done to the flowers and they can be visited again and the process repeated. As the fauna of flowers may change to some extent as the flowers age, it is a good idea to visit flowering trees and shrubs periodically so as to gain a representative sample of the species present.

Collecting from fruit

Fruit are similar to buds in many ways in that most of the insects associated with them are larval forms living within the fruit; many of them enter the soil to pupate. Fruit should be picked in various stages of development and stored in gauze-covered glass containers, away from damp and excessive heat and with a little clean soil or sand in the bottom. Adult insects emerging from the fruit or soil can be collected. It is important to make sure that the insects collected were not in the soil in the first place and it is advisable to heat the soil strongly in an oven to kill any insects present before placing it in the container.

Emergence boxes

Many insects are attracted to light and it is a good idea to have some boxes of various sizes, which can be made light-tight, with a hole drilled into one side, near the top, just big enough to take the mouth of a 2.5 cm glass vial. There must be no gap between the vial and the hole in the box. Material from which insects are expected to emerge is placed in the box and the box closed, making sure that there are no escape routes, even for small species. To increase the efficiency of this emergence box a white sloping board can be attached inside the box, sloping from just below the mouth of the vial to the floor of the box at a steep incline. The mouth of the vial should not protrude from the wall of the box on the inner side. The emerging insects will be attracted to the light coming into the box through the glass vial. The emergence box is best placed with the vial facing a window so that sufficient light enters the box so as to attract the insects. It should not be placed in direct sun as the vial will become overheated. When insects are no longer appearing in the vial the box should be opened, when it will probably be found that some specimens have remained in the box. If the box is larger than about 15 by 15 by 15 cm, more than one collecting vial can be provided. The emergence box can be used for any material in which insects have to complete their development before emerging.

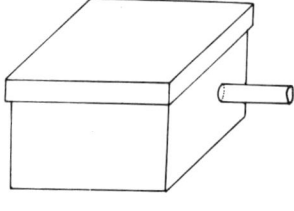

A simple emergence box.

Collecting from galls

Galls are abnormal developments of plant tissue caused by insects and the causative insect is found within the tissues of the gall. They can be found on almost any part of a plant and many of them are characteristic for the insect species. Some assume large and beautiful shapes, but most of them are small, in the form of bulges, swellings, or other distortions. The galls are produced by the plant in response to stimulation of some kind by the insect and removal of the gall from the plant before the insect is in an advanced stage of development often causes the death of the insect. In order to obtain the adult insects, therefore, it is better to collect galls towards the end of their development. Galls can be collected and put in an emergence box or gauze-covered jars. Insects of several orders are gall formers, but the formers are frequently parasitised by other insects and there are even parasites of the parasites (hyperparasites) as well as predators and insects just using the galls as a home (inquilines). A complex of species of insects is associated with most galls. Not all galls are caused by insects; many are caused by mites or fungi.

Collecting from bark

The bark of trees and shrubs harbours a specialised community of insect species. Many of these are small and many of the larger species are coloured so as to resemble their background and are difficult to see. Bark insects should be sought for individually; many of the species obtained by using a beating tray will actually be inhabitants of the bark of twigs or small branches rather than of the foliage. As bark insects are easier to see when they move it is a help to run the hand slowly up and down the bark, holding the hand a few centimetres from the surface of the bark. The crevices of bark should be inspected for insects hiding there and loose bark can be lifted. In doing so, do not rip the bark, but lift it gently in sections or strips. Some insects will remain where they are, others will fall to the ground or scuttle under adjacent sections of bark. Remember to inspect the undersides of the pieces of bark as you remove them.

Many insects hibernate beneath bark although they are not found in that situation at other times of the year. In some cases large numbers of such species can be collected as they mass together during hibernation.

Collecting at roots

Insects which live at the roots of plants have to be dug out. In the case of large plants, roots can either be followed down from the surface or the finer rootlets reached by digging a little way from the main stem; in the case of small plants deep shovels-ful of earth and plants should be lifted up and placed on a sheet of newspaper, brown paper, or opaque white plastic sheeting; the soil can be broken down and the roots examined. By the time the insects are found they may have moved or been shaken from their original positions at the roots and insects move away from the roots when not actually feeding. Also, the soil itself will contain large numbers of insects which are not necessarily associated directly with the roots present. Most of the insects at roots are not particularly active; they can be collected by hand or by forceps if large enough, otherwise by aspirator or camel-hair brush. The latter is less likely to pick up soil and debris and with a little practice the collector can become quite adept at singling out specimens.

Collecting from leaf litter

The leaf litter which accumulates under trees, particularly in damp localities, usually supports a wealth of insects mainly of small forms which can find food and shelter between the pieces of debris. Many of them have moved up from the upper layers of the soil and when disturbed retreat there quickly, disappearing between soil particles and under pieces of leaf.

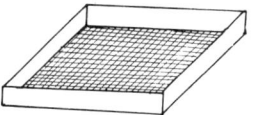

Simple sieve for sifting leaf litter.

Apart from searching amongst the decaying leaves and picking out the larger specimens there are three main ways of obtaining material from leaf litter. The first method is to pick up handfuls of leaf litter and throw it onto a sheet of white paper, cloth or opaque plastic sheeting; there it can be picked over and specimens leaving the mass of vegetable matter can be seen more easily against the white background. The second method is to use a sieve. This should be made of wood or light metal about 30–50 cm square and about 15 cm deep. The mesh, for general purposes, should be about eight to the 2.5 cm. Leaf litter is scooped up and put into the sieve which is then agitated over a white cloth, or sheet of opaque white plastic, preferably about a metre square.

The third and most efficient method for collecting leaf litter insects is by means of the 'Berlese funnel'. This is an item of equipment which cannot usually be carried into the field, although portable types have been described and of which various models differ from one another in detail. Basically, the Berlese funnel consists of either a metal funnel with a very smooth internal surface and with a mesh screen across it near the wider end or a dish with a mesh bottom. On this mesh is spread the material from which the insects are to be collected and above this is suspended a gentle source of heat. An electric light bulb is quite adequate if a reflector is arranged above it to direct the heat onto the leaf litter. The lower end of the funnel is inserted into the mouth of a container partially filled with liquid preservative (usually alcohol) or the dish is placed over a tray of liquid e.g. water with a little detergent added. The alcohol container should be corked with a hole bored through the centre of the cork just wide enough to take the end of the funnel. Any gap between the end of the funnel and the mouth of the container should be stuffed with cotton wool. As the leaf litter in the top of the funnel dries out, the insects move downwards finally ending up in the alcohol or liquid in the tray. Drying out of the litter may take several days and the catch can be removed periodically over that time. A Berlese funnel will give the best catch of litter insects, especially the smaller species. Material for funnel treatment can be transported in brown paper bags but should not be allowed to become heated by lying in the sun between collection and Berlese funnel treatment. Most leaf litter insects die when exposed to light and heat.

A simple form of Berlese funnel.

Collecting from birds' nests

Birds' nests are usually ignored by general collectors but they form an interesting habitat and frequently contain insects. The fauna usually varies with the age of the nest and according to whether it is occupied by birds. Nests can be dismembered in the field over a white sheet, after the most conspicuous specimens have been removed, or the whole nest can be brought home in a bag for later inspection. It is important to identify the species of bird concerned, and to make sure that the nest is not in use.

Collecting from dead wood

As soon as a plant dies, changes begin to take place in its tissues. The changes in the soft parts of plants are rapid and the communities of insects associated with them change rapidly. In the case of dead trees the wood decomposes at a slower rate and is colonised by many long-lived as well as short-lived species. Much of the decomposition of wood is performed by bacteria or fungi and there are insects found in decaying wood which are interested in these organisms and not the wood itself. Also, of course, the decaying bark, and the spaces below it, harbours an insect fauna of its own. As the wood decays the ground below it becomes rich in decaying vegetable matter and many species which feed on such material are attracted there.

For collecting insects from dead wood a hatchet, large chisel or a stout, pointed knife is useful; hatchets with a spike at the opposite side of the head from the blade are obtainable and these are quite useful. It is not necessary to have a heavy instrument. Chisels, knives and hatchets should always have the blades covered when not in use and they should *never* be handled carelessly in the field. When collecting from logs the bark, if present, should be inspected thoroughly, both on the surface and in the crevices. The bark should then be removed piece by piece. Some insects will remain where they are or move slowly; others will dash for cover. Collect the slower specimens first, remembering where the escapees went; they can then be extracted after the easier captures have been made. When the log itself has been thoroughly worked externally it should be turned over if possible. The last operation on the log consists of slowly taking it to pieces so that the various borers and other insects which live within it can be extracted. By the end of the operation the log will have been thoroughly broken down. If operations are not continued to this point the log, or its remnants, should be rolled back to its original position.

Collecting from fungus

Many species of flies, beetles and thrips are exclusively fungus dwellers and collections of mushrooms, toadstools and bracket fungi will yield many species. The fungi can be collected, preferably in various stages of development, including decaying material, and placed in glass containers, with a fine gauze covering. An emergence box will yield adults of fungus dwellers. Larval forms can be found inside the fungus.

Collecting from soil

The collection of insects from the soil is carried out in much the same way as collecting from leaf litter. The Berlese funnel is by far the most efficient for general collecting, although there have been a tremendous variety of techniques developed for extraction of soil insects, some of them highly specialised and aimed at extracting certain species only for research on pests and some of extreme complexity requiring expensive equipment.

Collecting from under stones

Turning over stones of almost any size will yield insects. Stones should be replaced to encourage repopulation. Samples of soil from below stones, scooped up after the larger specimens have been taken, can be put through a Berlese funnel.

Collecting from rocks

Large rocks, especially those in damp situations, have a fauna all of their own. Many insects settle on rocks to bask in the sun; these can be taken by sweeping a net across the rock a little above the surface. The lichen and moss which occurs on rocks (and elsewhere) will often yield species not found in other situations; after inspection, samples can be removed for Berlese funnel treatment or sieving over a white sheet. It is best to remove the lichen or moss by sliding a strong flat-bladed knife across the rock, cutting the material close to the surface. If moss is to be taken away it should be kept cool and damp until inspected as many moss insects are very susceptible to desiccation.

Collecting from dung

The droppings of animals, especially mammals, are an attractant for insects as soon as they are voided. Flies are usually the first to arrive and lay eggs on animal droppings. As the droppings dry out and are changed in other ways, so other insect colonists arrive, some to feed on the dung itself, others to feed on those species already present. The upper surface of a mass of dung from a large animal, such as a cow, hardens and dries but the lower layers retain moisture for quite a long time and such masses of dung can be easily dismembered in search of adult and larval insects; flies and beetles of several families will be found to be the most common inhabitants. The soil beneath and around dung should be inspected as many dung feeders complete their development below the dung. The emergence box is useful for collecting dung insects; a little soil in which to pupate is required for some species. Dung can be brought home in a plastic or thick-woven cloth bag. Plastic bags are fairly cheap and can be disposed of or they can be washed quite easily and re-used.

Collecting from carcasses

What has been said of collecting from dung applies also to collecting from carcasses. The sequence of colonising species is even more pronounced in carcasses than in dung and a greater number of species can be found. As it is not usually feasible to bring home a whole carcass, a sample of the rotting flesh can be taken and specimens taken from the rest of the carcass added to it. An outdoor location for the emergence box or other container is advisable owing to the smell of the decaying flesh. The horns of animals, if left untreated in any way, are usually colonised by moths in due course.

Collecting aquatic insects

In many ways the aquatic environment is a repetition of the terrestrial one with water replacing air. In consequence of this similarity there are a great many separate habitats to be investigated and various specialised techniques have been developed for collecting water insects.

The most important item for the general collector is the dip or drag net of which there are various forms. The ring may be semicircular instead of round, the side of the ring opposite the handle being a diameter of the circle. As a dip net is subjected to rough usage it should be made of fine material but protected by an apron of stronger material such as canvas or drill. Some water nets are triangular in shape, the handle being attached at one of the angles. A more elaborate drag net has a metal tube sewn into a circular opening at the bottom of the bag. The tube contains a filter of wire mesh and for convenience can have a terminal metal ring with a cap which can be unscrewed.

Dip and drag nets are used by dragging them through water weeds and across the bottoms of ponds and streams. The contents can be shaken out onto a white enamel tray and searched for specimens. After the violent disturbance of being banged about in the net many insects lie still or cling to the debris and these should be given an opportunity to become active again in the tray before the debris is discarded. Caddisfly larvae often hide within

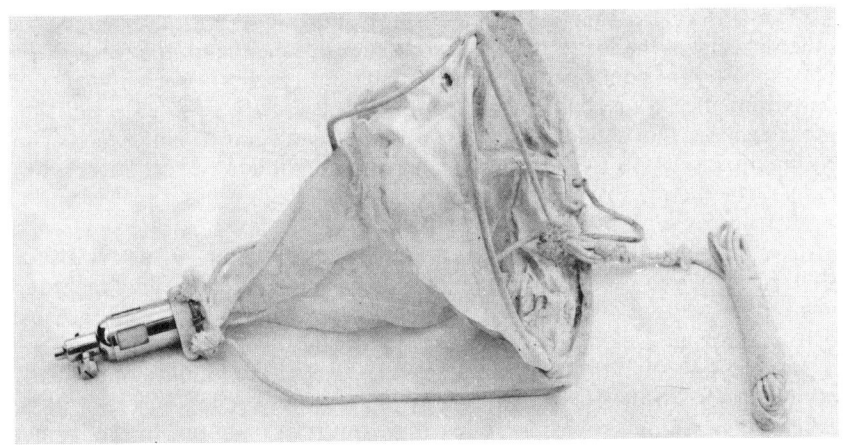

A dip or drag net with a metal container at the end in which the catch accumulates.

their cases for some time and are very difficult to see until they move about. It is often well worthwhile to tip the contents of many draggings into a bucket or other large container of pond water and return home with this 'concentrated' catch. It can then be tipped into an aquarium and the insects collected at leisure. As well as dragging through vegetation and open water the nets should be used to stir up the bottom mud, using the ring of the net whilst moving the net slowly forwards and against the current.

Masses of aquatic vegetation can be pulled up and inspected in a white tray. Many aquatic species have nymphal and larval forms that live on or under the rocks or logs at the bottom of the water. These can be turned over and inspected, while a net is held, mouth towards the rock, just a little downstream. Some of the insects disturbed but not seen will be swept into the net. Fresh water sponges and encrusting algae harbour the larvae of some insects and surface insects can be caught individually by net. Insects in wet sand and mudbanks alongside water can often be collected by throwing handfuls of sand or mud into the water. The sand or mud will sink, leaving the insects on the water from where they can be collected.

Collecting parasitic insects
Insects have exploited all manner of food supplies, including the use of other animals as food. In addition to using other animals and insects as prey many are parasites. The fleas, mosquitoes and other biting insects feed on warm blooded animals; biting lice also occur on birds and mammals and the sucking lice on mammals. An animal which lives upon another is called a parasite, the one on which it lives is called the host. The easiest way of obtaining parasitic insects is to search the hosts, especially immediately after death. Fleas leave the host body as soon as it begins to cool after death. The nymphal forms of lice are found on the hosts with the adults but the larvae of fleas are to be found in places where animal debris is found such as in lying-up places, nests, or in the dust of houses. Lice and fleas can be collected by parting the hairs or feathers and dabbing the insect with a drop of alcohol on a camel-hair brush. This will usually immobilise it.

If a body of a host animal is found it can be placed in a closed plastic bag for transport. Parasites will often be found free in the bag after a while, but not all of them will leave the host, which should be inspected thoroughly. Great care should be taken to ensure that specimens from different host individuals are not mixed nor parasites allowed to wander from one host to another.

Many insects live through their developmental stages within the bodies of other insects; at the same time the normal development and functions of the host species are continued. When the parasite reaches a certain point of development, e.g. just before its pupation, it leaves the host, usually killing it in the process. During this development the parasite may be quite selective in eating those parts of the host body which are not vital to its life. This type of parasite (sometimes referred to as a parasitoid, as it eventually does kill its host, whereas a parasite does not normally kill its host!) is often obtained instead of the adult when attempts are made to rear the host species as, for example, when caterpillars or other larvae collected in the wild are reared in captivity. The eggs of the parasite have already been laid on or in the host before it was collected. The most usual parasites obtained in this way are members of the fly family Tachinidae or species of the families of parasitic Hymenoptera. These insects are very common and many species will be taken by sweeping as well as by rearing from host material. In the former case, of course, the host species will not be known whereas in reared material it will be. In some cases thousands of small hymenopterous parasites can develop in one host caterpillar. The smallest parasites complete their development within the eggs of other insects! All stages of insects may be attacked by parasites, the egg, nymph, larva or pupa frequently having its own complement of parasites. Field collection of insects in various stages of development will yield many parasites if the hosts are kept alive. If parasite collection is intended, the container in which the hosts are reared should be insect proof, as many of the parasites are small and will escape from, say, a loose-mesh cover over a rearing container.

Collecting insects in traps

In essence, trapping insects consists of two processes, that of attracting the insect by some means and, having attracted it, of preventing its escape. The most obvious attractant for night-flying insects is light. Inspection of street lamps and shop windows, especially those with fluorescent lights, will prove productive. A similar lamp on a porch with a white wall or white sheet behind it will attract many species when conditions are suitable. Mains power is often not available in the best collecting places. Motor car headlamps shining on a white sheet will attract insects.

It is now possible to obtain portable low-voltage fluorescent lamps which will operate from a car battery. These bulbs emit light of wavelengths differing from those of ordinary incandescent bulbs and tend to be more attractive to insects generally. High-voltage mercury vapour bulbs some-times require special devices to control the rate at which they warm up; others can be used directly from the mains supply or from a 240 volt generator. Some emit rays which can be harmful to the eyes, others do not; it is therefore very important to check on these points with the manufacturer or retailer at the time of purchase of the bulb. Insects attracted to the lamp will often circle it or fly about in its vicinity, some will settle on a white sheet suspended by the lamp, others will settle nearby.

There are several designs of insect light trap on sale by natural history equipment suppliers, varying in complexity and cost, but mostly incorporat-ing a mercury vapour bulb. Most of these traps consist of a light source, below which is a funnel down which the insects fall, after striking the lamp, into a killing jar or dark box from which they cannot escape. A simple, easily constructed form of such a lamp is illustrated. This is adequate for general collecting. All automatic traps which enable moths to enter suffer from the same disadvantage, that is, there is a tendency for the scales from the bodies and wings of the moths to come off and stick to other specimens or make the inside of the receptacle dirty. Mercury vapour traps are extremely popular with collectors, especially moth collectors, and much has been written regarding every aspect of their operation and efficiency. It should be pointed out that many insects are not attracted to these lamps and that completely representative collections of even the moths in a particular district cannot be obtained by their use alone.

Suction traps consist of a suction fan above a funnel leading to a killing jar or container of liquid preservative below the fan. Their main function is to take a sample of the small flying insects such as aphids, thrips, psocids, etc. which pass over the trap. Their range is short and they are not of great use to the general collector. A simple form of aerial trap consists of a large screen, smeared with 'Tanglefoot', a substance to which the insects stick when they are blown against it. 'Tanglefoot' can also be smeared on cloth or brown paper which is then tied around a tree trunk in a band or series of bands. This will trap flying insects as well as insects which wander onto the 'Tanglefoot'. One disadvantage of this method is that the insects are not easy to dislodge and are sometimes damaged in the process. Many insects will take refuge behind bands of canvas or cloth tied around tree trunks. These can be removed periodically and the insects collected.

For trapping ground-dwelling insects, a jam jar can be sunk into the ground with its mouth level with the ground surface. Many insects, especially beetles, fall into this and cannot escape owing to the smoothness of the side walls of the jar. Some preservative can be placed in the jar.

A few traps depend on colour for their attractiveness. A shallow tray, about 3 cm deep and as large as practicable, can be painted yellow on the inside. This is filled with water and a little detergent and is then placed in the open, preferably a few feet above the ground. Insects, especially aphids and other small flying species, will be attracted to it and can be taken from the water. The detergent helps to wet the insects more thoroughly and so prevent their escape.

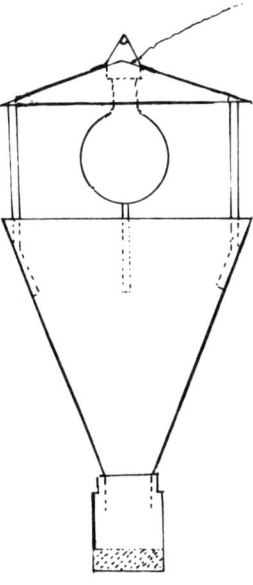

A design for a simple light trap.

Baiting for insects

Many insects will be attracted by baits alone. Meat is a good attractant for flies; sweet substances, such as jam or honey will attract members of various orders. Ripe or rotting fruit will attract flies and many beetles.

Moths are often attracted by 'sugaring', that is, to baits which are smeared or painted on tree trunks, fence posts etc. There are many recipes for sugaring mixtures in use and one of those which is widely used is given in the appendix.

In this chapter we have covered some of the collecting equipment and techniques used in general collecting. The ingenuity and requirements of specialised collectors have led to an infinite array of entomological equipment and techniques being designed and devised. Details of these can be found in the literature cited at the end of the book and the reader will probably be able to adapt or improve on these or even invent new devices to suit particular requirements.

It should be remembered that the more automatic a collecting technique is, the more specimens you are likely to get with less effort, but at the same time, you will learn less about the lives of the insects themselves unless you increase your study of the living insect. Many of the specimens caught by automatic collection belong to the same species; by hand-picking the catch this is avoided.

CHAPTER 4

PRESERVATION, MOUNTING AND STORAGE OF SPECIMENS

W HEN AN INSECT DIES chemical changes immediately take place in its body. These are later followed by the processes of decay which lead to the breakdown of the internal tissues. Having collected the insects and killed them the next task is to treat them so that they are preserved, permanently if possible, for later examination and study or for housing in the display cabinet. The fact that the insect cuticle is resistant to decay is a great help in our attempts to preserve the form of the insect after death. Insects are amongst the few groups of animals the form of which can be maintained in a dried out condition. It is, of course, important that the preserved specimens resemble the living insect as closely as possible; this is the object of any preserving technique.

There are two main types of preservation used for insect specimens. One consists, essentially, of letting the insects dry out after death, the other consists of immersing and storing them permanently in one of many available preserving fluids. The former technique allows the internal tissues to dry out and decay and the internal organs of a dry-preserved specimen are not, therefore, preserved for later study but only the external skeletal cuticle is retained. The liquid preservatives tend to preserve also the internal structures, with varying degrees of efficiency. Dry preservation tends to retain colour in more or less natural shades (but not all colours are retained) while liquid preservation tends to destroy colours or to alter them considerably. It is very important to bear in mind the use to which the specimens are to be put later when deciding on the method of preservation to be adopted for any specimen. In general, if only the external skeletal form and structures are required, dry preservation should be used; if it is intended to investigate the soft tissues of the body, liquid preservation is necessary; this is also necessary for soft-bodied insects in which the body becomes distorted by drying. The ultimate form of preservation to which specimens are to be subjected should be remembered in the field; specimens destined for preservation in liquid should not be allowed to dry out, even for a short while, after death. They should be put in liquid preservative in the field. We shall deal with dry and liquid preservation separately below and shall also discuss general techniques used in microscope slide preparation as some of the smaller insects should be prepared in that way for adequate study. At the end of this chapter will be found a synopsis of the generally used methods suited to each order of insects.

MOUNTING AND DRY PRESERVATION

Insects which are to be preserved and mounted dry may have just been killed or they may have been collected days or even weeks before and will have dried out. The first stage in their preparation will be pinning and setting but in order to do this the specimens must be supple and not dry and brittle. If the specimens have just been killed they will be in the right state, if they have been allowed to dry out they will need to be 'relaxed'.

Relaxing

Relaxing is the process of softening and making the specimen more supple and is carried out in a relaxing jar or tin. This is any jar or tin which can be made airtight and of sufficient capacity to take at least several large insects. A container of about 20 cm × 12 cm × 8 cm would be sufficient. Onto the bottom of this should be put about 2.5 cm of fine, clean sand, saturated with water (with no free water being allowed to remain) and to which has been added a quantity of carbolic acid (phenol). Over the wet sand is placed several layers of paper fitting to the edge of the container. Insects to be relaxed are placed in the closed box. The insects can be left in their papers or between layers of tissue paper but relaxing may take a little longer using this method than if they are not enclosed. Hard and fast rules cannot be laid down for the period of relaxing but one to seven days is adequate for most specimens. The relaxing jar should not be allowed to stand in the sun or become warm. For very hard, dry specimens which would be damaged by a prolonged period in the relaxing jar, Barber's Fluid (see Appendix) may be used. The insect can be immersed in this, or a drop or two of the fluid can be placed on the insect. It should not be used for Lepidoptera.

Desiccator used as a relaxing jar.

Instead of using a flat tin or jar for the relaxing jar, a desiccator can be used. This is a glass container, obtainable in various sizes, used in chemical laboratories. When used for relaxing insects the lower half of the dish is filled with wet sand and a little carbolic acid, fine wire gauze is placed across the jar resting on the internal shelf and the upper half of the jar is used to hold the specimens to be relaxed. A little petroleum jelly should be smeared on the ground glass edge of the dish to ensure airtight contact with the lid. In order to open and close the dish the lid should be slid sideways.

Pins and pinning

Adult insects are usually mounted on entomological stainless steel pins, the pins being passed through some hard part of the body (usually the thorax) from above. Ordinary pins should not be used for this purpose as the body fluids of the specimens quickly cause corrosion of the pins. The pins are obtainable in a variety of sizes. There are two series of sizes, the 'English' and 'Continental', and pins of various qualities are on the market. In each series the pins are numbered according to their sizes and thicknesses. Manufacturers have not standardised this numbering and so we shall refer here to measurements instead of size numbers. It is preferable to standardise in the use of pins of about 35–40 mm in length, varying the thickness according to the size of specimen. The thicknesses available run through a wide range and the diameters available are approximately as follows: 0.34 mm, 0.38 mm, 0.43 mm, 0.46 mm, 0.50 mm, 0.59 mm, 0.66 mm, 0.70 mm, 0.78 mm. Thicker and thinner pins are also available but if an insect is so small as to need a pin of less than 0.34 mm it is preferable to make a 'double mount' (described later) as a long pin of less than that diameter is very flexible and difficult to manage without danger of damage to the specimen. Clearly, it is not necessary to have on hand pins of all these thicknesses; three or four thicknesses, spread out over the range will suffice. Such a series would have, say, thicknesses of: 0.34 mm, 0.38 mm, 0.43 mm and 0.50 mm. A few of a larger size for bigger species could also be on hand with advantage.

Insects with a reasonably hard cuticle are pinned simply by passing a mounting pin through the body from above. The positions vary according to the order of insects but all pins should be passed through the body vertically so that when the point of the pin is stuck into a piece of cork, the body of the insect is horizontal. Most insects are pinned through the thorax, in the midline between the fore wings. There are some exceptions. Orthoptera are pinned through the prothorax, Hemiptera through the scutellum (triangular, posterior extension of the back of the mesothorax), Coleoptera are pinned through the right elytron near the front adjacent to the midline.

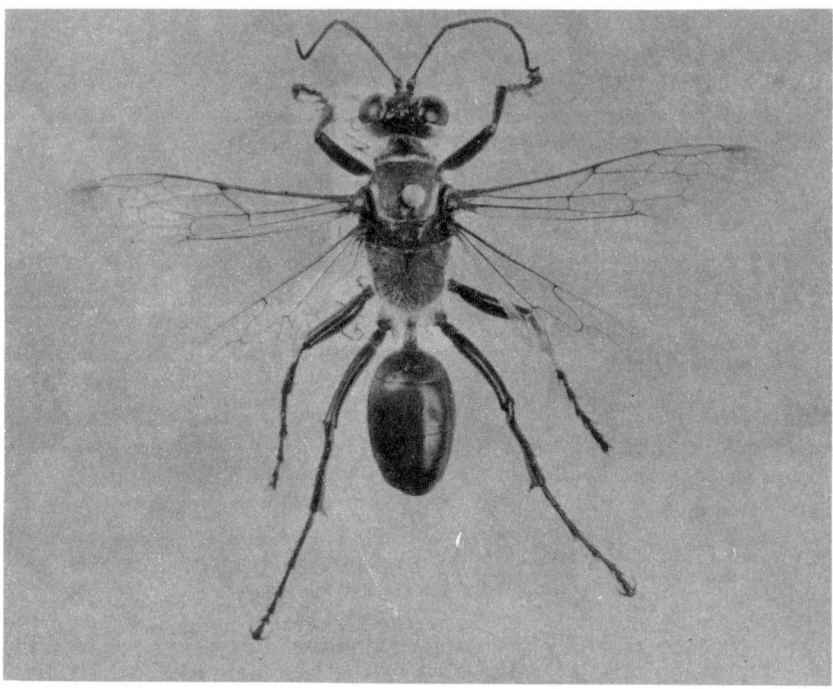

Position of mounting pin—wasp.

*Position of mounting
pin—moth.*

*Position of mounting
pin—beetle.*

*Position of mounting
pin—bug.*

The insect should rest about three quarters of the way up the pin. This will allow enough room above it for the pin to be handled easily by fingers or forceps and if all specimens are pinned at a uniform height on the pins the collection will have a neater appearance. In cases where several specimens of one species are available some may be pinned with the ventral side uppermost; this is a common practice with Lepidoptera.

Pinning forceps

Once an insect has been mounted on a pin it should not be moved more than necessary by hand, but the pin should be gripped by pinning forceps. These are strong forceps about 10–12 cm in length with broad, curved, milled prongs. Pins should be gripped firmly near the point; this will minimise bending and springing of the pin and reduce the chances of damaging the specimen.

Double mounting

When specimens are too small to be mounted on an ordinary entomological pin, they may be double-mounted. In this type of mounting the specimen is attached to a piece of card, pith, polyporus or other material and the pin is then pushed through this.

Double mount with minuten pins

The specimen may be mounted on a headless, stainless steel minuten pin, again about three-quarters of the way up the pin. Minutens are shorter and finer than mounting pins, being from 10–20 mm in length and ranging from about 0.14 mm to about 0.30 mm in thickness. The best lengths are of 10 to 15 mm and the thinner diameters are to be preferred. The minuten is usually pushed through the insect from above, but in some groups, e.g. small flies, the pin may be inserted through the thorax from one side. A piece of 'polyporus' (obtainable from dealers in strips of various lengths and about 3 mm × 3 mm in cross section) is cut to a suitable length, usually about 10–15 mm. In the case of an insect pinned on a minuten from above the point of the minuten is pushed through the polyporus near one end of the strip, the point of the minuten being allowed to protrude a little through it. Through a point near the other end of the polyporus strip an ordinary mounting pin is pushed and the polyporus slid up the pin so that the specimen is at the same height as other specimens in the collection which are mounted directly on pins. Many collectors are dogmatic on the question of which way the specimen should face when double mounted and in which direction the polyporus should lie in relation to the mounting pin. This is not a matter on which to have fixed ideas; the best arrangement is that which will make for easiest examination of as many sides of the specimen as possible.

Double mount with card points

Instead of using minuten pins the insect can be mounted on a card point. This is a triangular piece of good quality white card (Bristol board is often used) cut to a size depending on the insect to be mounted; the usual length is about 1 cm with a base of about 3 mm. The point opposite the base of the triangle should be bent a little and the specimen glued by its side or lower surface to this point with a *small* droplet of *water-soluble* glue. If several specimens of the same species are available some can be glued the other way up. An ordinary mounting pin is pushed through the cardboard triangle near the base and the card point raised up the pin to a height about three-quarters of the way up.

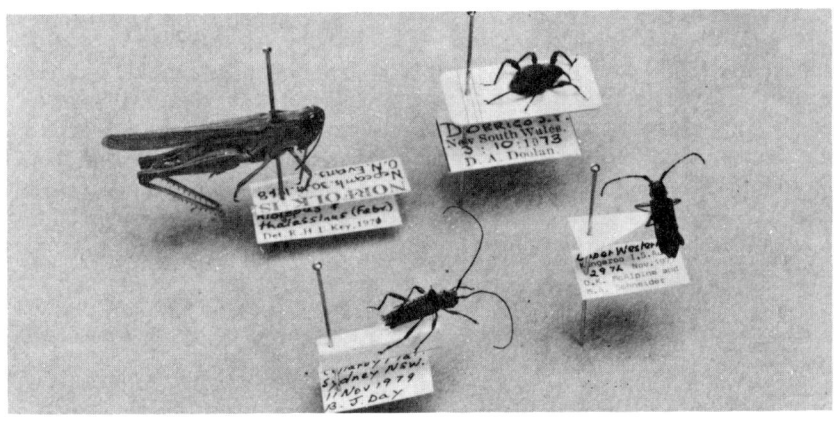

Various ways of 'double-mounting' insect specimens (the beetles) compared with a normal mount (the grasshopper).

65

Mounting on card

Many collectors glue their specimens onto the upper surface of a rectangular piece of card and then pin the card in the same way as in the case of a card point. This method is not generally recommended as the underside of the insect (or upper side if the insect is glued back down) cannot be inspected without removing the insect from the card.

Setting

If the wings and appendages of a relaxed insect are placed in a particular position and held there whilst the insect is allowed to dry out the arrangement will be maintained permanently by the dry specimen. This is called 'setting'. Insects should always be in a relaxed condition when pinned so that any adjustments which have to be made so as to make antennae, limbs or wings more easily inspected can be made at the same time. In some groups of insects it is essential for the wings to be spread so that details of colour and venation can be seen. In others, the legs or antennae carry important features used in identification and must be clearly visible. Some collectors are very meticulous in their setting and spend considerable time on it but, like all things, setting can be overdone and one wonders whether time spent on setting minute detail on a specimen is worthwhile if examination of the specimen is not made proportionately easier. A well-set specimen is one which shows you what you need to see. In specimens intended only for exhibition the aesthetic interest may be greater than the scientific interest and the time spent may then be considered worthwhile.

Many insects require little in the way of setting beyond a little spreading of antennae and legs. This can be done with a fine camel hair brush or needles after pinning. The abdomen of elongated insects may droop; this can be overcome by inserting the mounting pin into a piece of cork sheeting and supporting the abdomen in its natural position by means of a strip of stiff card on another pin or the insect can be pinned to a vertical piece of cork so that the abdomen hangs downwards. Crossed pins can be used to support the abdomen. When the insect has dried out the supports can be removed and the abdomen will remain in position. Insects are brittle after drying.

Setting board

Insects which require setting with their wings out, such as butterflies, moths, dragonflies, grasshoppers, katydids, cockroaches, mantids and lacewings etc. are set on a setting board. A series of boards of different sizes will be needed. As with most items of entomological equipment, setting boards vary in details of construction. A typical board is made up of a plywood base board, to which is glued a layer of softer material such as cork or polystyrene to take the point of the mounting pin and two side strips, on which to spread the wings. The depth from top of side strips to top of base board should be about equal to the distance up the pin at which the insect is to be set. Total board width and groove width will depend on insect size so a series of boards should be made. For a board of 5 cm width the space between the side strips should be about 6 mm; for a board of 3 cm the space should be about 5 mm. Smaller and bigger boards should be of approximately the proportions given above but the active collector will gradually acquire a series of boards to suit his needs.

Some setting boards are made simply from wooden blocks with grooves of various widths along the middle and with the board on either side of the groove covered with cork sheeting. The design described above will be found to be very easy to use as the pins which will hold the insect can be pushed in

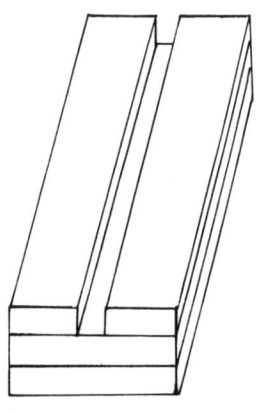

A simple setting board.

quite easily. Some boards have the wing-supporting slats sloping slightly away from the central groove but it is preferable to have these horizontal. For setting minute insects, such as very small moths, a miniature board made along similar lines can be used.

Using the setting board

To set an insect on a board, the following procedure should be adopted. Pin the insect vertically through the thorax with a mounting pin of suitable size. Choose a board with a groove width a little wider than the body width of the insect. Insert the pin into the cork in middle of the groove and push it down until the point is firmly embedded in the lower cork strip. Make sure that the pin is vertical. Push the insect up or down the pin so that its body is accommodated in the groove with the base of the wings level with the top of the cork on the wing-supporting slats. With boards of the sizes suggested, the insect will be about three-quarters of the way up a 30–40 mm pin. Using a long, fine pin, a needle mounted in a wooden handle or a pair of forceps, spread the wings out on the wing-supporting slats. The wings will probably not remain in the positions in which they are put but will tend to hinge back towards their natural resting position. Place a narrow strip of paper across the wings some distance from the body and pin it to the board well in front of the front margin of the fore wing on one side and secure it with another pin at the other end behind the hind wing. Then, with a pin or needle pull the fore wing forward, sliding it gently along under the paper strip, until its hind margin is at right angles to the body. Holding it in this position, place a pin in a position immediately in front of the front margin of the fore wing and pin the strip at this point securely to the board. This will usually hold the fore wing in position. With the fore wing held firmly, pull the hind wing forward until its front margin comes to lie just under the hind margin of the fore wing. Move the back pin forwards and insert it so as to secure the strip just behind the hind wing. The strip will now be pinned down over the wings so as to hold them securely in position. Carry out the same procedure with the wings of the opposite side, making sure that the insect is symmetrically arranged and that its body lies straight along the groove. Making sure that the wings are held in their positions, pin a broader strip of paper over the wings using several pins placed close to the margins of the wings and placed so as to make the paper hold the wings close to the board so that wing movement is impossible. There is no need to use the special insect mounting pins for any purpose other than actually mounting the insect. If so desired, the legs and antennae can be arranged, if they can be reached without upsetting the specimen; a fine needle with a small hook at the end is useful for this. The specimen is now set and is ready for drying.

A series of specimens, from top to bottom, at various stages in the setting process.

Many groups of insects are set with their wings out, some are set with the wings of only one side spread out, those of the other side being allowed to remain in their natural resting position. The usual methods adopted for each order are given in the synopsis of treatments for each order (see page 81).

Drying

Before the animal can be placed in its permanent storage place, it must be dry. This is best done by placing the setting board in a container away from dust and insect or other pests. A large closed (but not airtight) box with some naphthalene flakes in the bottom is suitable, or the board can be placed in a cupboard. Wherever the board is placed, it should have adequate air circulation. Freshly pinned specimens which do not need setting can be pinned to sheets of cork or polystyrene foam and similarly protected until dry. Drying may take anything from a week to several weeks depending on the humidity of the air and the state and size of the specimen. When dry the strips of paper can be carefully removed from the setting boards and the mounting pins, with the specimens attached, can be lifted out. It must be remembered that dried insects are very brittle and fragile; they should be moved only by holding the mounting pin below the specimen by means of a pair of forceps (preferably pinning forceps). At no stage during setting and drying should an insect specimen be separated from its appropriate data label. Whilst on the setting board the label (temporary or permanent) should be pinned next to it and the label should accompany it whilst it is drying.

Labelling

Labelling is one of the most important operations in making a collection and yet it is surprising how frequently one finds collections with inadequately labelled specimens or, even worse, specimens without labels at all. The need for recording collection data at the time of collection has already been stressed.

The permanent label for material which has been preserved dry is mounted on the pin below the specimen. The label should be of good quality stiff white card; it will have to last for many years. The label should be as small as practicable but there is no particular virtue in having the label and writing on it so small that a magnifying glass is required to read it! For general purposes about 12 mm × 8 mm is satisfactory. The label should be placed about half way up the pin. If the label is too close to the specimen it will not be easy to read without handling the specimen. The side of the label with the writing should be uppermost. Some collectors are jealous of letting others know where they collected their specimens and attach the label with writing downwards. In order to read the information the specimen has then to be lifted up and turned over. This leads to unnecessary handling of the specimen with the attendant danger of damage. In any case, such a habit indicates lack of appreciation of the real functions of a collection. Data labels should be written in permanent ink, such as India ink. Any ink which has the slightest tendency to fade over the years should be avoided. A mapping pen or one of the modern drawing pens designed to take India ink should be used. The finer the better, as small lettering is more easily executed with a fine pen. A little practice will enable the collector to prepare labels in small writing quite well. Needless to say, labels should always be neatly written so that there can be no possibility of misreading the information. When a collector is operating in one or several localities frequently over a long period it will save time if he has printed labels giving the locality and his name, with a space for the insertion of the data which varies, such as habitat and date. If a light trap is operated labels for the specimens taken can be printed. The type size for data labels should be 6 point or less. If there is too much information to fit onto one label a second label can be placed on the pin a little below the first. If extensive collecting from one habitat or plant species is anticipated a number of labels with that information alone can be printed and one placed above or below the other data label.

STORAGE OF DRIED SPECIMENS

Prior to pinning and setting dried specimens can be stored in the containers in which they had been brought from the field. If they are to be left for any length of time they should either be allowed to dry out, which will necessitate later relaxing, or a few crystals of chlorocresol should be put into the containers before sealing them so as to be airtight.

When a specimen has been pinned, it is ready for permanent storage. If the specimen has been identified it will probably be placed in position in the collection; if not, it will be stored temporarily pending identification. The permanent dry-mounted collection can be housed in store boxes or in specially constructed insect cabinets; temporary storage will usually be in store boxes.

Store boxes

A store box is a well-constructed wooden box with a close-fitting, hinged wooden lid, the floor of the box being lined with a sheet of cork, the whole of which is lined with white paper. Various sizes are available but it is essential that they be deep enough to take the longer type of pins when the box is shut. In many models the lid is as deep as the box and is also cork lined, thus doubling the capacity of the box; in such cases the distance between the surface of the cork lining of the floor and the lining of the lid should be about 8 cm. This will prevent the heads of the mounting pins from touching one another. Common sizes of store boxes are about 25 cm × 20 cm, 37 cm × 25 cm and 50 cm × 30 cm. The box should be made so that it is insect proof and dust proof.

Store boxes come in various sizes and shapes.

The pins should be pushed gently but firmly into the cork lining of the box, the pin being gripped low down by forceps. Specimens should be arranged neatly in rows in the store box with enough room between them to allow manipulation of the pinning forceps. Naphthalene flakes (moth balls) should be placed in the box (see under 'Pest and Mould Prevention', page 73). It is also sometimes possible to adapt other boxes, such as cigar boxes or other domestic boxes by lining their floors with cork, balsa wood, plastic foam or any other material in which mounting pins will be firmly held. The disadvantage of such boxes is that it is virtually impossible to make them insect proof and the frequency of inspection necessary to ensure that the specimens are not being attacked by pests is inconvenient. When pests do get in they are difficult to eradicate from such boxes.

Smaller sized store boxes (often called posting boxes) can be obtained. These are useful for holding a few specimens at a time and after being suitably packed can be used for sending specimens through the post.

Insect storage cabinets

Insect storage cabinets are amongst the most expensive items of equipment required for entomological work; for this reason many private collectors house their specimens in a large series of store boxes as the monetary outlay is then spread. A good insect cabinet is usually a fine example of the cabinet maker's art and requires considerable skill in manufacture but many home-made cabinets are found which serve their purpose very well. Second hand cabinets are occasionally offered for sale by collectors or furniture dealers.

An insect cabinet consists of a series of glass-topped drawers, each drawer is lined on the bottom with cork sheet and lined overall with white paper. A cabinet with more than ten drawers is difficult to move. It may have a hinged or sliding door; many designs have a lift-off door but a door is not essential

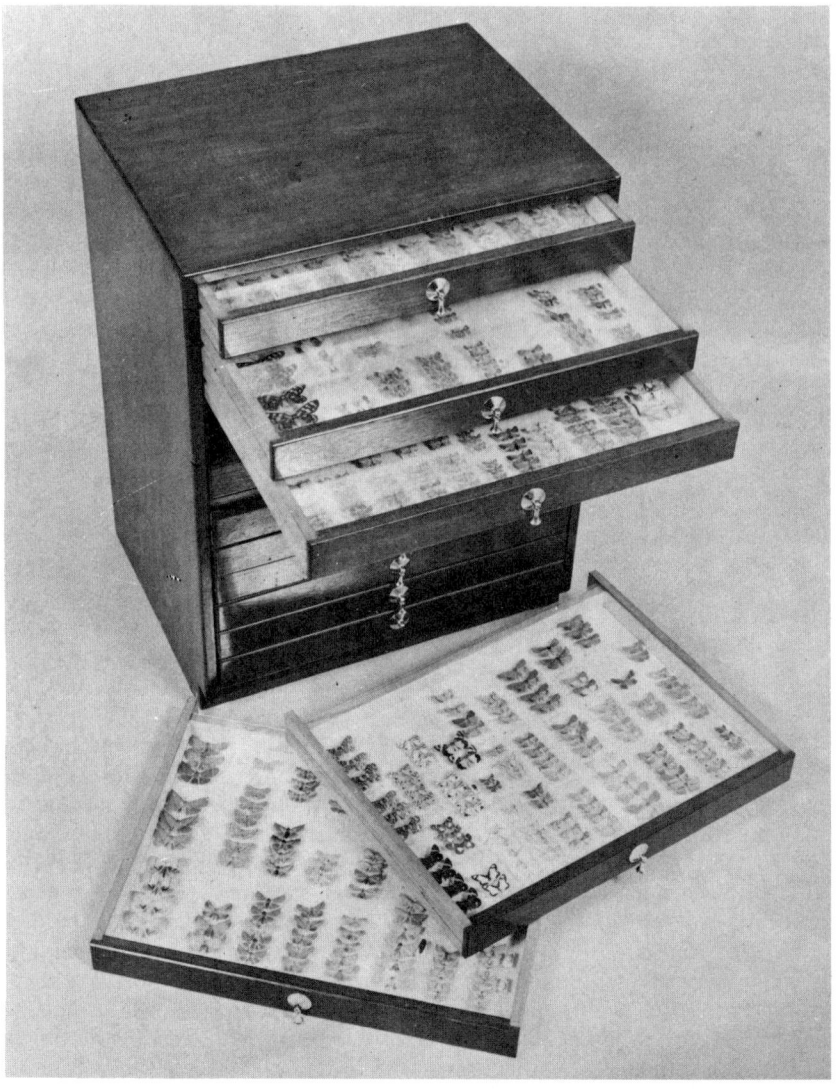

An insect storage cabinet.

and can be a nuisance when room-space is limited. Cabinets and drawers should be of hardwood. Soft woods frequently exude resins which have adverse effects on the specimens and cause the drawers to become stiff on their runners. The weight of drawers is quite great, particularly in larger cabinets, and the cabinet should be constructed so as to take the weight without warping. It is important to make sure that the floor on which the cabinet stands is absolutely flat otherwise the cabinet will gradually become distorted and the drawers will then not slide smoothly.

The size of drawer varies, the smallest being about 40 cm × 40 cm; some of the larger ones are about 60 cm × 45 cm. A rectangular drawer is probably preferable to a square one in the larger dimensions as the longest side can then be across the drawer and it does not have to be pulled too far out of the cabinet. The depth of the drawer should be such that the long pins can be accommodated without their heads touching the glass top. The drawers must be made insect proof and the lids must be tight-fitting; also, the drawers should be made so that they are interchangeable throughout the cabinet or cabinets. This is very important in an expanding collection and if drawers are not interchangeable a considerable amount of moving of specimens from drawer to drawer will be necessary when additional specimens are to be inserted in a crowded drawer. Around the inside of the drawer, a little away from the wall there should be a second wall, leaving a space of about 7 mm between it and the main wall. The inner wall should not be as high as the side of the drawer. The channel between the two walls is kept filled with flake naphthalene or other insect deterrent and by having the inner wall lower than the outer a gap is left between the inner wall and the lid of the drawer which allows the naphthalene vapour to enter the drawer space which houses the specimens. Details of a suitable type of drawer and cabinet are illustrated. Cabinets and drawers made of metal are now obtainable; they have the advantage of being machine made and therefore identical series

Details of drawer construction.

71

should be obtainable to ensure interchangeability of drawers. The drawers are usually heavy and the bottoms, being of a single sheet of metal may sometimes tend to buckle. In warm, damp climates where there are big differences in day and night temperatures the danger of moisture condensation on the metal surfaces may be greater than in wooden drawers.

Arranging the specimens

Dried specimens are usually arranged in the drawers or store boxes according to one of the accepted schemes of classification. Species are arranged in columns starting at the back of the left hand side of a drawer. A 60 cm drawer can have about six columns of moderately sized insects across its width, that is, a column of 10 cm is quite convenient. Columns are varied according to species size. Males are sometimes placed in one row and females in another, especially with Lepidoptera and in species in which sexual differences are obvious. The column may be marked by making a light pencil line on the paper lining or by putting a pin at the front and rear of the box between each column and stretching a length of fine black cotton between them, the cotton lying on the floor of the drawer. The names of the genera and species are written in fairly large letters on labels pinned to the floor of the drawer. They are usually arranged as follows: in the back left corner, at the head of the first column, the name of the family is placed; following this in the column comes a label bearing the name of the genus; then follow the specimens of the first species in that genus after which is placed a label bearing the name of the species, pinned to the bottom of the drawer. The next species follows, without repetition of the generic name in full, and its name also follows the specimens. In this way the generic name appears before the specimens of all species belonging to it and the specific name for each species follows the group of specimens belonging to that species. Some space should be left for the insertion of additional species and on the bottom right hand corner of the drawer a space can be left for related specimens which cannot be identified immediately.

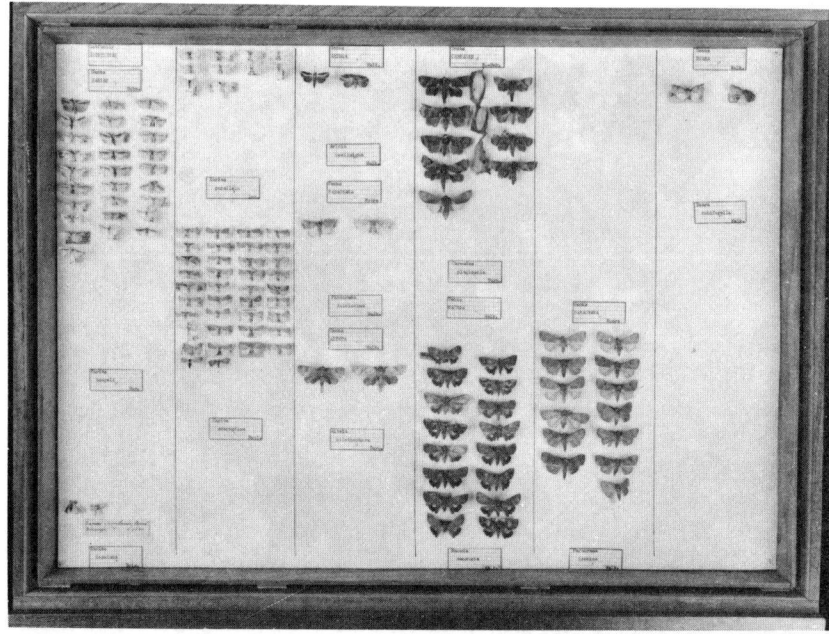

Specimens arranged in a drawer.

Each order and family is indicated by a label placed at the beginning of each category and it is usual for the front of the drawer to have a card holder into which can be slipped a card indicating the general nature of the contents of the drawer.

There has been a recent tendency to omit the cork lining from the bottom of the drawers and to house the specimens in open-topped cardboard trays placed within the drawer, the bottoms of the trays being of cork. In this system trays of standard lengths and widths are used so that a given number of trays will fit across the length and width of the drawer. In fact, a column of trays replaces each column in the previous system. Each species is placed in a tray of its own and the name of the species is placed on a label attached to the vertical end of the tray. By using trays of various lengths but equal widths, species with many or only a few representatives in the collection can be accommodated. Variation in size of specimen is allowed for by having various widths of trays. The advantages of this system lie in the fact that the specimens of each species can be handled as a unit. It can, however, be fairly expensive in that the cost of the trays may be relatively high.

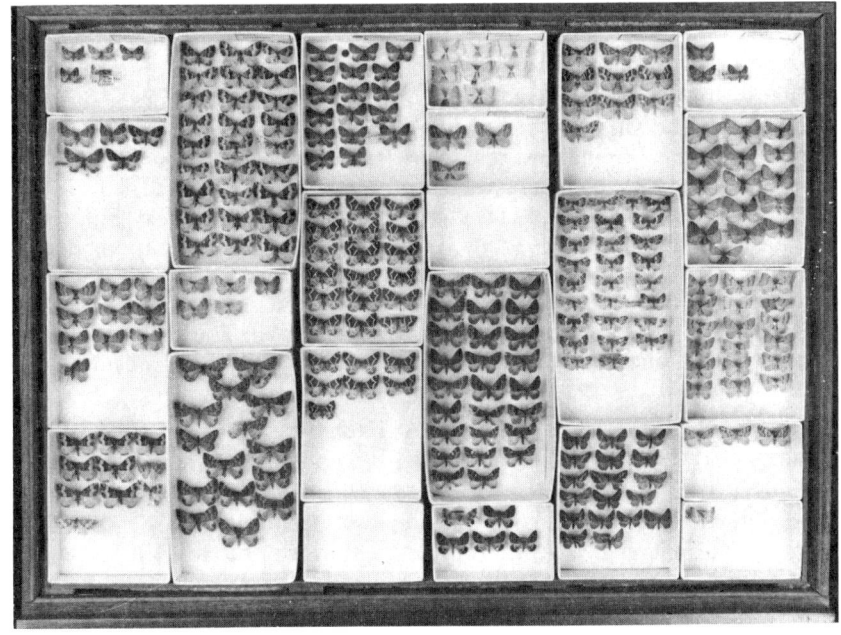

Specimens arranged using the 'unit tray' system.

Specimens of most groups which are preserved dry are pinned as described in this chapter. The fragile craneflies (family Tipulidae) can be stored dry in transparent envelopes with a data label enclosed and the envelopes filed in containers, much as cards are filed in a card index.

PEST AND MOULD PREVENTION

Dried specimens are very liable to damage by mould and mite or insect pests such as museum beetles and psocids. To prevent mould formation a little creosote can be placed in the store boxes or cabinets; small non-spill, glass containers on pins can be obtained for this purpose from dealers in natural history supplies. The boxes or drawers can be painted with a pest and mould deterrent (see Appendix) or a wad of cotton wool, dipped in warm melted phenol and allowed to cool, can be pinned in the corner of a store box. The crystalline phenol can be melted by putting the glass container in which it is held into a bowl of hot water. During temporary storage of material which is not dry, chlorocresol crystals should be placed in the containers. To prevent mould in the relaxing jars, phenol or trichlorphenol should be added to the wet sand.

The first line of defence against pests is to make sure that store boxes and cabinet drawers have close joints and tight-fitting lids. They should always be provided with an insect deterrent such as naphthalene. Do not sprinkle naphthalene loosely in a store box amongst the specimens as this will shift about when the box is moved and damage specimens. Instead, a small gauze or cloth bag of about 5 cm × 3.5 cm or a little larger, can be filled with naphthalene or moth balls and pinned securely in one corner of the box. In cabinet drawers the channel around the edge should always contain naphthalene flakes and all boxes and drawers should be inspected for pests at regular intervals. Naphthalene is volatile and may disappear quite rapidly in some climates, especially if the boxes or drawers are not well made.

If pests are found it will be necessary to fumigate the collection. The boxes or drawers should be opened and placed in an airtight container large enough to take them with a certain amount of free space to permit air circulation. The most useful fumigant is carbon bisulphide. This is a volatile liquid, foul smelling and extremely inflammable; in certain concentrations it is explosive. When it is being used all flames should be extinguished and smoking completely prohibited. A quantity of the liquid should be put in a shallow open bowl placed above the articles to be fumigated. The container being used for fumigating should be sealed and left for about twenty-four hours. It is preferable to carry out the fumigation out of doors or in an outside building. If a single drawer is to be fumigated, a quantity of paradichlorbenzene can be put in before sealing. Under such enclosed conditions the specimens will be effectively fumigated in a few days. It is advisable to repeat fumigation in a week or two. The pest and mould deterrent mentioned earlier can also be used as a fumigating agent.

DEGREASING

Some insects, especially the larger Lepidoptera and some Coleoptera, contain a great deal of fat in the body which gradually seeps out and discolours the specimen and drawer. A greasy specimen can be degreased by immersing it, still on its pin, in carbon tetrachloride or benzol. The liquid will become discoloured as the fat is extracted and should be changed if this is excessive. When the fat has been removed the specimen should be allowed to dry before being returned to the collection. Carbon tetrachloride and benzol are fairly volatile and drying is, therefore, quite rapid.

Ether, being a fat solvent, can also be used. This is more volatile than benzol or carbon tetrachloride and is used in the same way.

PRESERVATION IN LIQUIDS

Many insects can be preserved dry because the insect cuticle is relatively resistant to decay and is sufficently rigid to retain its form after the decay of the soft internal tissues. Those forms with a soft cuticle, such as most larvae and many of the smaller soft-bodied adults, should be preserved in liquid preservatives. The liquids penetrate the tissues and prevent decay; the outward form of the insect, as well as its internal soft organs, is preserved to a greater or lesser degree. Many insects are too small to be preserved dry owing to the difficulty of manipulating them in the mounting processes; these too, can be preserved in liquid. Coleoptera and Diptera are not preserved in liquid apart from exceptional cases.

Preservation of specimens, adults as well as larvae, in liquid is becoming more widely used for insects. The collection does not present as attractive an appearance as a neat, well-set collection of dried specimens but many of the finer structures suffer less distortion in liquid preservation than they do from drying and specimens are less likely to be damaged during handling. Insects in liquid can be handled in large numbers. It is quicker and easier to store large series of specimens of a species in liquid than it is to mount and set each one individually; also, should internal structures need to be examined they are preserved, at least to some extent.

Liquid preservatives

Only the common preservatives are dealt with here, an indication of which groups should be preserved in liquid is given in the synopsis at the end of this chapter. Formulae for these will be found in the Appendix. For general purposes 70–75 per cent ethyl alcohol is used. Specimens can be dropped directly into the fluid which will kill and preserve them. If the strength falls below 30 per cent its preservative power declines. In the case of caterpillars, beetle grubs and fly maggots it is advisable to kill and sterilise the larvae first by placing them in boiling water and leaving them to cool or by using a penetrating killing agent. After this they can be placed in other fluids, for example, alcohol, for permanent preservation. Larvae with very thin cuticle shrink as alcohol extracts water from the tissues. In such cases, the larva should be passed through strengths of alcohol of 25 per cent and 50 per cent before being put into the final strength of alcohol. It should be left in each concentration for at least two hours. This will extract the water more slowly and result in less distortion.

Alcohol has the disadvantage that it tends to harden the insect, but this is not serious as in most cases the insect is not set in any way.

For many purposes ordinary industrial (colourless) methylated spirits is adequate. It is usually supplied at a concentration of about 90 per cent or a little stronger. The colourless type only should be used. When diluted with water the purple-coloured type is liable to precipitate a white substance which coats the specimens and makes the liquid somewhat opaque. Formalin, which is a general preservative used for many animals can be used for insects as a 2–4 per cent solution. This preservative has poor penetrating powers and destroys colour to a greater degree than alcohol. It also hardens the specimens considerably and is used only for soft specimens. Commercial formalin is a 40 per cent solution of formaldehyde in water. Lacto-alcohol is a useful preservative if the specimens are being stored prior to mounting on a microscope slide in gum chloral. If lacto-alcohol is used the container cannot be closed by a cork as it will gradually be destroyed by the preservative. Data labels in lacto-alcohol should be written in soft pencil.

Instead of using hot water for killing larvae a solution known as KAA can be used as the fixative. The insect is dropped alive into the solution and when it has become fully distended it is removed and preserved in 95 per cent alcohol.

Storage

Specimens in alcohol are stored in glass vials. A limited selection of sizes should be used and as vials can be obtained in almost any length and any diameter it is convenient to select four or five sizes and use these where possible. A range of such sizes would be, say: 7.5 cm × 2.5 cm, 6.5 cm ×

Glass jar containing small specimens in alcohol in small vials.

1.8 cm, 6.0 cm × 1.2 cm, 5.0 cm × 1 cm. One or other of these sizes will accommodate most species. A tight wad of cotton wool, dipped in preservative to drive out the air, is used to close the vial and the vial is then turned up side down in a wide mouthed jar, filled with the same preservative, large enough to take a number of vials and fitted with an airtight lid. Preserving jars with a glass lid secured by a screw-on ring and sealed by a rubber washer are excellent. (Note that rubber is damaged by Carnoy's and Pampl's fluid). A layer of cotton wool can be placed in the bottom of the jar and pressed well down. This will decrease the chances of the vials rattling against the bottom of the jar and being broken. Each large jar might contain representatives of a single species (if much material is available) or the species of one genus may be placed in one jar. Should only a few species in a family be in the collection, one jar might be used to accommodate these. The manner in which the vials are distributed through the jars will depend on the nature and amount of the material available.

The jars should, like the dry collection, be arranged in sequence in accordance with an accepted scheme of classification and they should be arranged according to the same classification as the dry collection. The jars can be arranged in rows on shelves, preferably in a cupboard or, if this is not possible, the shelves should have a curtain to prevent fading of specimens.

Some collectors cork the individual vials and stand these on shelves or place them in racks instead of putting the vials in larger preservative-filled jars. The time spent in checking the vials periodically is considerable in any but the smallest collection and it is easier to check alcohol levels in jars than in vials. The vials may become completely dry and the specimens ruined in a short time. In order to prevent this a little glycerin is added to the alcohol. This will sometimes prevent the drying out of the specimens for long enough to enable the danger to be realised.

When the vials are immersed in preservative in larger jars it is only necessary to top up the larger jars from time to time to ensure that the level does not fall. This is a much quicker and easier task than inspecting and topping each vial, particularly in a large collection.

It is now possible to obtain glass vials with plastic push-in stoppers. The stoppers are resistant to most components of preservatives but sample tests with the preservative to be used should be carried out with each type of stopper used. When using stoppers of any kind, whether cork or plastic, it is preferable to fill the bottle with preservative so as to reduce the amount of air as much as possible.

Labelling

In order to save space in the dry collection the labels attached to the pins should not be too big but labels in vials can be quite large and each should have all the usual information. They can be almost as long and nearly as wide as the vial.

The information should be written in India ink, which must be completely dry before the label is put into the preservative. If it is not dry some of the black particles of which the ink is composed will drift off leaving the writing pale and the loose particles will adhere to the specimens. Information should not be written on a label attached to the outside of the vial, nor should it be written on the vial or stopper. Ink should be tested before use as some kinds of India ink fade in alcohol.

The contents of each jar of vials, whether the jar be devoted to a species, genus or family, can be indicated by a large label, written in India ink, placed within the jar. It can be run along the inside of the glass with the writing facing outwards where it will be visible without having to delve amongst the vials.

PRESERVATION ON MICROSCOPE SLIDES

Minute insects can be kept in the smaller sizes of glass vials but their small size makes it necessary to have a compound microscope to look at them if the details of their structure are to be seen. Many of the tiny insects are objects of great beauty. Special techniques are sometimes necessary in making microscope mounts but many specimens can be mounted satisfactorily by following a few simple rules.

Slide making

In principle, mounting an insect on a glass slide for examination under the microscope consists of several stages. The specimen is first prepared by removing the internal soft parts and leaving only the harder skeletal cuticle. Then all the water is removed by immersion in a dehydrating fluid. This is then removed by other reagents and the specimen is finally placed on a glass microscope slide in a substance (the mounting medium—usually resinous) which is permanent and into which it is sealed by means of a thin piece of glass called a cover slip. Thin or pale specimens may be stained during the process of preparation. The liquids in which the specimen is placed are usually held during use in 3.5 cm watch glasses or in hollowed out square glass blocks (called staining blocks). Specimens are transferred from one watch glass to another by means of a needle mounted in a wooden handle. It is important to take as little fluid as possible with the specimen when moving it from one watch glass to another as the quantities of the fluids are quite small (a few drops in each watch glass) and they will dilute each other if they are moved with the specimen from glass to glass. The microscope slide itself is a piece of glass, obtainable in various thicknesses, measuring 7.5 cm × 2.5 cm; the usual thickness is about 1 mm. The cover slip is a round or square piece of glass, obtainable in a range of sizes and thicknesses. The square ones range from about 12 mm to 2.5 cm square and round ones from a diameter of about 6 mm to 2.5 cm. Rectangular cover slips of various sizes are also obtainable. The thicknesses vary and each is numbered, No. 0 being the thinnest; they range from about 0.08–0.30 mm. For general purposes No. 1 cover slips (0.17 mm) are adequate with about 18 mm diameter or about 18 mm square.

There are many media which can be used for mounting insects, some more satisfactory than others. We shall deal here with the processes leading to the use of three of these, namely, Canada balsam, euparal and gum chloral.

Canada balsam mount

In order to soften the cuticle and remove the soft internal tissues, the specimen is placed for a few minutes in a 10 per cent solution of potassium hydroxide (caustic potash). If the insect has a very thick cuticle it may be left for longer in the potash. Warming the potash will speed up the softening process and with hard structures it may be necessary to boil the specimen in potash. Warming can best be done by placing the specimen in a small vial containing potash and warming this in a larger receptacle of water which is then heated. Boiling can be done in a test tube although care should be taken to hold the tube away from the flame so that the boiling is gentle or the specimen will be shot out of the tube as the potash suddenly bubbles. The flame should not be applied to the bottom of the tube, but about three-quarters of the way up the liquid. The mouth of a hot potash tube should

Method of heating a specimen in caustic potash in a water bath.

always be pointed away from the operator and care should be taken not to spill the substance, especially on skin or clothes.

The specimen should not be left in the potassium hydroxide for longer than it takes to soften the cuticle and liquify the body contents as colour will be removed all the time the treatment is continued. Prolonged treatment will make the specimen unnecessarily pale. After this maceration in potash the specimen is transferred to water for a few minutes to remove the potash. A trace of acetic acid added to the water will assist in neutralising the potash.

As Canada balsam is not soluble in water it is necessary to remove from the specimen all trace of water. This is done by transferring the specimen from the water to 30, 50, 70, 90 and 100 per cent alcohol in that order, leaving it in each for a minute or two in the case of small insects and for about ten minutes for larger specimens. A second treatment with 100 per cent alcohol is advisable. Absolute (100 per cent) alcohol is difficult to maintain in that state as it readily absorbs water from the atmosphere. The watch glass should therefore be covered when it is being used. It is, however, essential to remove all traces of water from the specimen prior to Canada balsam mounting. When this has been done the specimen is transferred to xylol. (If any water is left in the specimen it will go cloudy in xylol instead of becoming transparent.) When it has become transparent through the clearing action of the xylol it is ready for mounting in the balsam.

Canada balsam is a resinous substance dissolved in xylol, hence the xylol from the specimen will be miscible with the Canada balsam. The next step is to take a microscope slide and, after making sure that it is quite clean, place a small drop of Canada balsam in the centre of it. The amount of balsam required can only be discovered by experience. Canada balsam bottles are specially shaped with a cover and provided with a glass rod for dripping the balsam onto the slide. The balsam should be about the consistency of treacle; it can be thinned with xylol if it is too thick. The specimen is transferred to the drop of balsam and arranged in the required position. A *clean* cover slip is then placed over it by lowering the cover slip so that one edge of it meets the slide to one side of the drop of mountant (Canada balsam) and the specimen. The cover slip is supported by a needle and gently lowered over the specimen and allowed to settle down of its own accord. Bubble formation should be avoided. A label, about 2.0 cm × 2.0 cm, is stuck to the slide at one end and on it the usual collecting data is written. It is important to use labels which are gummed so that they will adhere to the glass slide indefinitely.

The Canada balsam is fluid and if the slide is tilted the cover slip will move. Until the balsam is dry enough to hold the cover slip, therefore, the slide should be kept horizontal. In due course the balsam will dry out and the mount will be permanent, the balsam eventually hardening completely. There is then no need to take steps to prevent the movement of the cover slip. Should it be desired to stain the specimen, reference should be made to one of the books suggested for details of the techniques involved at the end of this book. Staining is usually carried out before or at some point during the extraction of water by alcohol, depending on whether the stain being used is water-soluble or alcohol-soluble.

Specially shaped Canada balsam bottle, with wide cap and glass dropper rod.

Euparal mount

To make a mount in the mounting medium known as Euparal, the same procedure is adopted as for mounting in Canada balsam but the specimen can be taken directly from 95 per cent alcohol to the mountant without intermediate treatment. Xylol treatment, of course, is not necessary with Euparal as the mountant itself has a clearing action. If it is desired to thin the

78

mountant, a special diluent is available for this which can be purchased from the suppliers or it can be thinned with a little butyl alcohol. Euparal mounts are permanent, the mountant eventually becoming hard.

Gum chloral mount

To make a gum chloral mount the specimen is kept in potash for as short a time as possible in which to effect the necessary removal of soft tissue. With very small specimens potash treatment can be omitted. It is then placed in water for a few minutes. From this it may be placed in the mountant on the slide. After the cover slip has been placed in position the slide should be warmed gently and then left to cool. The mountant will be partially dried and set in about two weeks but it is then necessary to carry out a process known as 'ringing'. This consists of making a watertight seal around the edge of the cover slip. To do this, any excess mountant on the slide outside the rim of the cover slip should be carefully scraped off. A layer of one of the many ringing compounds available should be painted around the edge of the cover slip and the adjacent part of the slide. An excellent ringing medium for use with gum chloral is 'Glyceel'. This can be purchased from dealers; if too thick it can be diluted with butyl acetate.

In most cases the specimens will be fairly thin and the cover slip will lie more or less flat upon the microscope slide. If thicker objects are mounted it may be necessary to prop up the cover slip so that a thicker layer of mountant can be held below it. This can be done by placing four slivers of glass from a broken cover slip under the four corners of the cover slip, or a square frame of paper or thin card can be cut to fit under the cover slip to leave a cavity in the middle of the frame. Small metal circles (cells) are obtainable in various thicknesses for the purpose of raising cover slips.

Slide storage

When the mountant used does not become completely hard, it is necessary to have some form of storage for the slides in which they lie horizontally. In the case of completely hardened mountants, they can be stored on edge, either standing on end or on their sides. It is best to purchase a suitable type of cabinet; most of these consist of flat trays in which from a few to many slides lie side by side, the trays sliding into a wooden or metal cabinet one above the other or lying one upon the other in a box. The amount of storage space required will decide the type of cabinet to be used. As a general principle it is better to obtain a small unit first and, as the collection grows, additional units can be purchased or made. The use of small units makes expansion of the collection easier and reduces the amount of slide moving required as the collection grows.

Another type of cabinet consists of a box just wide enough to take a row or two of slides, with a series of slots on the inside of each side of the box. The slides are slipped into the slots and are stored on edge, but can be stored horizontally if the box is kept on end. Such boxes can be stored next to one another on shelves like books. Cardboard boxes containing slide holders are available from dealers in natural history supplies and are probably best for small collections.

Sending insects through the post

Most collecting excursions will be of short duration, but from time to time the collector may have the opportunity and pleasure of undertaking a longer trip or perhaps of being a member of a collecting expedition. Under normal circumstances a collector will have his specimens with him and bring them home himself. There are, however, times when it may be necessary to send specimens by post and at such times they will have to be adequately packed. A collector may wish to send specimens to other collectors either as a gift, for study, for identification or as part of some exchange arrangement.

Dried specimens in papers or between sheets of tissue will travel quite well if some crystals of chlorocresol are included to prevent mould in case the parcel becomes damp in transit. They can be packed in small tins or boxes, well wrapped. Single specimens can be sent in small boxes. The specimens should be packed so as to prevent movement in the parcel and the box in which they are held can be surrounded by a generous quantity of wood wool or cotton wool before being wrapped. Wrapping paper should be strong and securely tied with string or strips of adhesive paper.

If the insects are mounted they can be sent pinned in a store box. It is advisable to place ordinary pins around each specimen, close to the body, to prevent the insects from swivelling on their pins. Should one specimen in the box become loose it will knock pieces from its neighbours and these will fall free into the box and damage further specimens. All the specimens in a box can be ruined by one insect swivelling or falling from its pin. The mounting pins should be firmly embedded in the cork so that there is no danger of their coming loose. Again, one loose specimen can ruin them all. A bag of naphthalene balls should be securely pinned in the box or the box painted out with a pest deterrent mixture. A store box with pinned material should be packed in a bigger box or carton surrounded by a *thick* layer of cotton wool, wood shavings or equivalent material which will absorb shock. It is important that the store box be well free of contact all round with the containing box or carton so that the store box cannot be directly jarred. The container should be well wrapped and secured with string or adhesive paper.

Liquid-preserved material should be packed in sealed bottles so that there can be no leakage. Plastic stoppers, or corks dipped in molten wax after they have been placed in the bottle, will usually ensure this. Each vial should be wrapped in a piece of paper so that if they do break the contents of each is kept separate. A wad of cotton wool should be placed in the vial so as to gently hold the specimen still against the bottom of the bottle; excessive fluid movement during transit can cause damage to the specimens. The vials can be packed in small boxes so that they cannot move about and these boxes treated in the same way as store boxes.

Single vials can be wrapped in paper, then in several thicknesses of corrugated cardboard and then in brown paper. The address should be placed on the parcel and on a tie-on label attached to the securing string. The postage stamps should be on the tie-on label, not on the parcel itself. A useful device for sending single vials through the post consists of a rectangular balsa wood (or other wooden) block a little longer and wider than the vial. This has a hole bored into its end that is big enough to take a vial. A wad of cotton wool is pushed into the bottom of the hole, the vial is inserted and the hole closed at the end with a tight wad of cotton wool and the block wrapped in brown paper. Each parcel should be clearly and completely addressed, with a return address given in case of non-delivery or in case of query. If the parcel is going to a foreign country there may be formalities to be completed to satisfy quarantine and customs authorities that the material is neither harmful nor to be sold. There is not usually any difficulty in this regard if the contents are clearly stated to be dead and preserved specimens for scientific study and that they are not being sent for commercial purposes. If the recipient is associated with an educational or scientific institute of any sort use the address of the institute, if the recipient is in agreement, in preference to a private address. A record should be kept by the sender of all parcels and their contents with a note when they were sent and from which post office. In case of valuable specimens they can be registered or insured.

SYNOPSIS OF TREATMENTS FOR EACH ORDER

The following synopsis includes the generally-used techniques for dealing with the insects of each order. In the larger orders there are often families or sections of the order for which specialists have devised special techniques. These are not included as they usually apply to insects not collected by the general collector.

In this synopsis killing methods are marked D and W to indicate which is to be used if dry (D) or wet (W) preservation respectively is to follow. The preservation methods to follow in each case are then indicated by D or W according to whether the killing agent used is appropriate for dry or wet preservation. That is, if the specimen was killed by a method marked D it should be preserved by a method marked D. M indicates a microscope slide preparation of the whole insect, which is made from a liquid-preserved specimen. Where 'killing jar' is given as the killing agent it should be understood that any of the vapourous killing agents mentioned in Chapter 3 can be used; exceptions to this are mentioned and in some cases a specially recommended killing agent is mentioned separately. In each case an indication is given to which stages in the life cycle the technique is appropriate.

Insect eggs are preserved in 70–80 per cent alcohol.

1. ORDER THYSANURA

Killing: W Drop into 75% alcohol.
Preserving: W In 75% alcohol with a little glycerin added.
              M Mount in Euparal or Canada balsam.

2. ORDER DIPLURA

Killing: W Drop into 75% alcohol.
Preserving: W In 75% alcohol with a little glycerin.
              M Mount in Euparal.

3. ORDER PROTURA

Killing: W Drop into 75% alcohol.
Preserving: W In 75% alcohol with a little glycerin.
              M Mount in Euparal or Canada balsam

4. ORDER COLLEMBOLA

Killing: W Drop into 75% alcohol.
Preserving: W In 75% alcohol with a little glycerin.
              M Mount in Euparal.

5. ORDER EPHEMEROPTERA

Killing: D Killing jar (adults).
              W Drop into 70% alcohol (adults and nymphs).
Preserving: D 1. Mount on card (adults).
              D 2. Pin through centre of thorax and set with all wings spread (adults).
              D 3. Double mount, using minuten pin (adults, small species).
              W In 70% alcohol (adults and nymphs).

6. ORDER ODONATA

Killing: D 1. Killing jar (ethyl acetate not recommended) (adults and nymphs).

D 2. Inject ammonia (commercial strength) into thorax by hypodermic needle (adults).

D 3. Drop into nearly boiling water (nymphs).

W 1. Drop into 70% alcohol (nymphs).

W 2. Drop into Oudeman's Fluid (nymphs).

W 3. Drop in nearly boiling water (nymphs).

Preserving: D 1. Pin through centre of thorax and set adults with all wings spread. (If desired, before pinning, internal organs of abdomen can be removed by cutting along ventral side of abdomen from near base to near apex and extracting contents. Abdomen should then be stuffed with cotton wool dusted with borax).

D 2. Mount on card (nymphs).

W 1. In Oudeman's Fluid (if killed in that fluid) (nymphs).

W 2. In 70% alcohol (nymphs).

7. ORDER PLECOPTERA

Killing: D Killing jar (adults).

W Drop into 70% alcohol (adults and nymphs).

Preserving: D Pin through centre of thorax and set with all wings spread (adults).

W 1. In 70% alcohol (adults and nymphs).

W 2. In 2% formalin (adults and nymphs).

8. ORDER GRYLLOBLATTODEA

Killing: W Drop into 70% alcohol (adults and nymphs).

Preserving: W In 70% alcohol (adults and nymphs).

9. ORDER ORTHOPTERA

Killing: D 1. Killing jar (adults and nymphs).

D 2. Expose to fumes of benzol (adults and nymphs).

W 1. Drop into 70% alcohol (adults and nymphs).

W 2. Drop into nearly boiling water (adults and nymphs).

W 3. Expose to fumes of benzol (adults and nymphs).

Preserving: D 1. Pin through prothorax and set with wings of one side spread. Before pinning, stuff abdomen of large specimens as in Odonata (adults).

D 2. Pin through prothorax (nymphs).

D 3. Double mount on minutens (small species).

W In 70% alcohol with a little glycerin (adults and nymphs).

10. ORDER PHASMIDA

Killing: D Killing jar (adults and nymphs).

W Drop into 70% alcohol (adults and nymphs).

Preserving: D 1. Pin through mesothorax and set with wings of one side spread. (Stuffing of abdomen can be carried out as for Odonata) (adults).

D 2. Pin through mesothorax (adults and nymphs).

W In 70% alcohol (adults and nymphs).

11. ORDER DERMAPTERA

Killing: D 1. Killing jar (adults).
                D 2. Expose to fumes of benzol (adults).
                W Drop into 80% alcohol (adults and nymphs).
Preserving: D 1. Pin through right elytron and set with wings of left side
                     spread (adults).
                D 2. Double mount on minuten pins (smaller species).
                W In 80% alcohol with a little glycerin (adults and nymphs).

12. ORDER EMBIOPTERA

Killing: D Killing jar (adults).
                W Drop into 70% alcohol (adults and nymphs).
Preserving: D Pin through thorax and set with all wings spread (adults).
                W In 70% alcohol (adults and nymphs).

13. ORDER BLATTODEA

Killing: D 1. Killing jar (adults and large nymphs).
                D 2. Drop into near-boiling water (adults and nymphs).
                W 1. Drop into 70% alcohol (adults and nymphs).
                W 2. Drop into nearly boiling water (adults and nymphs).
Preserving: D 1. Pin through centre of thorax and set with wings of one side
                     spread (adults).
                D 2. Mount on card (small nymphs).
                W In 70% alcohol (adults and nymphs).

14. ORDER MANTODEA

Killing: D Killing jar.
                W Drop into 70% alcohol.
Preserving: D Pin through thorax and set with wings of one side spread.
                W In 70% alcohol (adults and nymphs).

15. ORDER ISOPTERA

Killing: W Drop into 75% alcohol (adults and nymphs).
Preserving: W In 75% alcohol (adults and nymphs).

16. ORDER ZORAPTERA

Killing: W Drop into 75% alcohol (adults and nymphs).
Preserving: W In 75% alcohol (adults and nymphs).
                M Mount in Euparal or Canada balsam (adults and nymphs).

17. ORDER PSOCOPTERA

Killing: W Drop into 75% alcohol (adults and nymphs).
Preserving: W In 75% alcohol (adults and nymphs).
                M Mount small species in Euparal or gum chloral (adults and
                    nymphs).

18. ORDER MALLOPHAGA

Killing: W Drop into 80% alcohol (adults and nymphs).
Preserving: W In 80% alcohol (adults and nymphs).
                M Mount in Euparal (adults and nymphs).

19. ORDER SIPHUNCULATA

Killing: W Drop into 70% alcohol (adults and nymphs).

Preserving: W In 70% alcohol (adults and nymphs).

M Mount in Euparal (adults and nymphs).

20. ORDER HEMIPTERA

Killing: D Killing jar (adults of Heteroptera and larger Homoptera).

W 1. Drop into 70% alcohol (adults and nymphs of all families).

W 2. Drop into lacto-alcohol (aphids and scale insects).

Preserving: D 1. Pin through scutellum and set with wings of one side spread (larger species, adults).

D 2. Pin through scutellum (smaller Heteroptera and Homoptera, adults and nymphs, nymphs of larger Heteroptera).

D 3. Double mount on minuten pins or card point (smaller species, adults and nymphs).

W 1. In 70% alcohol (adults and nymphs).

W 2. In lacto-alcohol (aphids, scale insects, adults and nymphs).

M Mount in gum chloral (aphids, scale insects, adults and nymphs).

21. ORDER THYSANOPTERA

Killing: W 1. Drop into 70% alcohol (adults and nymphs).

W 2. Drop into mixture of 50% methyl alcohol and 0.5% ethyl acetate (by volume). Preserve specimen within three days (adults and nymphs).

Preserving: W In 70% alcohol (adults and nymphs).

M Mount in Canada balsam.

22. ORDER NEUROPTERA

Killing: D 1. Killing jar (adults).

D 2. Chloroform (adults).

W 1. Drop into 70% alcohol (all stages).

W 2. Drop into Carnoy's Fluid (larvae).

Preserving: D 1. Pin through thorax and set with all wings spread (adults).

D 2. Mount on card (larvae).

W In 70% alcohol (all stages).

23. ORDER MECOPTERA

Killing: D Killing jar (adults).

W 1. Drop into 70% alcohol (all stages).

W 2. Drop into Carnoy's Fluid (larvae).

Preserving: D Pin through thorax and set with all wings spread (adults).

W In 70% alcohol (all stages).

24. ORDER LEPIDOPTERA

Killing: D 1. Killing jar (must be dry) (adults).

D 2. Kill with ammonia (adults of very small species).

D 3. Chloroform or ethyl acetate (larvae).

D 4. Inject with Carnoy's Fluid (pupae).

W 1. Drop into Kahle's Fluid (larvae).

W 2. Drop into boiling water (larvae and pupae).

W 3. Inject with Carnoy's Fluid (larvae and pupae).

Preserving: D 1. Pin through thorax and mount with wings of both sides spread (adults).

D 2. Mount on minutens, set on miniature setting board and double mount (very small species, adults).

D 3. Pin or mount on card (pupae).

W 1. In 70% alcohol away from light (from boiling water, Carnoy's Fluid or Kahle's Fluid) (larvae).

W 2. In 70% alcohol (after injection with Carnoy's Fluid or from boiling water) (pupae).

25. ORDER TRICHOPTERA

Killing: D Killing jar (adults).

W 1. Drop into Pampl's Fluid (adults, larvae and pupae).

W 2. Drop into Pampl's Fluid (adults, larvae and pupae).

Preserving: D 1. Pin through thorax and set with wings of both sides spread (adults).

D 2. Double mount on minutens, after setting (small species, adults).

W 1. In 70% alcohol (adults, larvae and pupae).

W 2. In 3% formalin (from Pampl's Fluid) (adults, larvae and pupae).

Note: Larvae can be preserved in their cases or the latter removed and mounted on card for dry preservation. It is preferable to retain the case in liquid with the larva.

26. ORDER DIPTERA

Killing: D 1. Chloroform (adults).

D 2. Killing jar (adults).

W 1. Drop into 80% alcohol (larvae, pupae).

W 2. Drop into KAA for two hours (larvae).

W 3. Drop into boiling water (larvae).

Preserving: D 1. Mount on cards, specimens glued on their sides (adult Tipulidae).

D 2. Mount on card points (small species, adults).

D 3. Pin through top of thorax (adults, larger species).

D 4. Double mount on minutens from above (adults of smaller species).

D 5. Pin through side of thorax with minutens, double mount (adults, small species).

D 6. Pin from below with minutens, double mount (adults, small species).

D 7. Store in cellophane envelopes (adult Tipulidae).

W 1. In 80% alcohol (from 80% alcohol or boiling water) (larvae and pupae).

W 2. In 95% alcohol from KAA (larvae).

M Mount in Canada balsam (small species, adults).

Note: Several special techniques are used for some families of this order; works listed below should be consulted for details.

27. ORDER SIPHONAPTERA

Killing: W Drop into 70% alcohol (adults, larvae and pupae).

Preserving: W In 70% alcohol (adults, larvae and pupae).

M Mount in Euparal (adults and larvae).

28. ORDER HYMENOPTERA

Killing: D Killing jar (adults).

W 1. Drop into 70% alcohol (adults and larvae).

W 2. Drop into Carnoy's Fluid (larvae and pupae).

Preserving: D 1. Pin through thorax and set with all wings spread (large adults).

D 2. Mount on card points (small adults).

D 3. Pin on minutens and double mount (adults).

W In 70% alcohol (from alcohol and Carnoy's Fluid) (adults, especially small species, larvae and pupae).

Note: Bees of all kinds are usually preserved dry with the mouthparts and legs set in an extended position.

29. ORDER COLEOPTERA

Killing: D 1. Killing jar (adults).

D 2. Ethyl acetate fumes (adults).

D 3. Drop into boiling water (adults).

W 1. Drop into Carnoy's Fluid (larvae).

W 2. Drop into 70% alcohol (larvae).

W 3. Drop into Pampl's Fluid (pupae).

Preserving: D 1. Pin through right elytron, near front (adults).

D 2. Mount on minuten pin, placing minuten as in D 1. then double mount (adults, smaller species).

D 3. Mount on cards (small species).

D 4. Mount on card points (adults, small species).

W 70% alcohol (from Pampl's Fluid, Carnoy's Fluid or alcohol) (larvae and pupae).

30. ORDER STREPSIPTERA

Killing: D Killing jar (winged males).

W 70% alcohol (males, females in their hosts).

Preserving: D Mount on card points (winged males).

W 70% alcohol (males, females in their hosts).

M Mount in Canada balsam or Euparal (adults).

CHAPTER 5

REARING INSECTS

T HE DRY SPECIMENS IN the cabinet or the preserved ones floating in alcohol represent far more than themselves; they represent what were once living insects. The specimen of the insect is merely the vehicle which carries its life, and its life is by far the most important and interesting part of the animal. Maintaining insects alive is an interesting and important part of entomological activity and only by keeping live specimens under observation can we come to know anything about the insects' place in the world. Insect rearing is carried out for various reasons and from a scale which involves only a few specimens to one in which thousands upon thousands are reared.

We have seen that most insects undergo metamorphosis; this means that there is more than one form which individuals of the species take during their lifetime. It is not quite correct, therefore, to think of the adult form as representing the 'species' any more than we would ignore children when considering what is a human being. To know fully what a species looks like we should really take account of the various immature stages and the only way of knowing these is to rear them where they can be observed and under conditions which can leave no doubt that a certain specimen is, in fact, the young of a certain adult.

The most obvious reason for rearing an insect is to obtain specimens of all the stages of the species for collections. In many cases collectors who are only interested in having specimens of the adults seek out mature or nearly mature larvae and nymphs and keep these until the adult emerges. In this way, of course, they can obtain more or less unblemished adult specimens for their collections although there is usually no greater scientific value in an unblemished specimen than in a naturally worn one. The information relating to a specimen is the important thing as long as the specimen is identifiable beyond doubt.

Rearing a species will give material on which to base descriptions and illustrations of the immature forms as well as providing a series of specimens of the various stages for the collection to which reference can be made later. A collection of insects in which there has been no attempt to include immature forms, or during the making of which immature forms have been ignored when available cannot be considered a 'good' collection. It is surprising how many collectors have reared thousands of specimens from larva to adult form and yet their extensive collections contain few, if any, preserved immature forms as though the immature forms had no place in the lives of the insects they collect. Happily, many of the younger generation of collectors are more realistic in this matter and are taking more interest in the immature stages.

Apart from merely obtaining specimens for collections and for anatomical or descriptive work, insects are reared with a view to finding out something of their life history; finding out how long they live in each stage, what they feed on at each stage and the many matters which fall within the scope of a general life history study. This type of information is as important as the specimen itself and it is usually necessary to have some general knowledge of the life history of a species before more detailed studies can be made.

In rearing insects it is usually adequate to have reasonably small numbers of specimens and with proper planning of observations quite a lot can be learnt from these. If work of a detailed nature is to be undertaken, such as studies in the physiology or of the behaviour of the species it may be necessary to have many specimens of known age, condition and sex. It may also be necessary to rear them so that material is available at certain seasons of the year. In such cases the life history of the insect would have to be known in detail before the rearing programme could be planned and the actual planning of the rearing programme would have to be detailed. Many species have been, and are being, reared for use in testing the effectiveness of insecticides to be used against pest species. Sometimes the actual pest species is reared in large numbers but more often species which are easily reared are used for routine testing. In such cases the species may be mass reared with continuous and successive cultures of the insects being set up so that a large and continuous supply of the insects of known age and condition are available. The insects are produced in thousands much as articles are produced in a factory; in fact, mass rearing laboratories resemble small factories in many ways. Mass rearing is also used in the production of parasites and predatory species for use in biological control work, that is, in the use of an insect species to combat a pest species or to control a noxious weed. The insects are produced in enormous numbers by mass rearing techniques and released at strategic points at an appropriate time in an effort to reduce the pest populations.

Only very few species, of course, ever come to be reared on this scale. By far the greater proportion of reared species are kept for study of the insects themselves and are, therefore, generally handled on a small scale.

It has been pointed out that there must be a great many species of insects which are not yet described; of those which have been, only a very small proportion have had their immature stages described and an even smaller proportion have had their life history investigated in even the simplest manner. On the other hand, a few species, mainly those of economic importance, such as pest species, or those of special value in research, such as the vinegar flies (*Drosophila*) in genetic research, have had their biology intensively studied. For those species which are well known and which are in constant use or under constant study, highly involved and complex rearing programmes and equipment are used based on detailed knowledge of the insects' life history and requirements for survival. For the vast majority of species, this knowledge is not available and, in any case, for rearing small numbers, simple apparatus can usually be devised. There is a great need for simple studies of living insects, for the elucidation of the salient features of their life histories. It is true that in order to rear an insect some knowledge of its life history is necessary but as insects of the same genus, and even family or order, often have broadly similar habits it is often possible to obtain some idea of what to expect and how to start out in the first place with a species which has not been reared before by reading how its relatives live. In order to be able to set about devising rearing methods a knowledge of some of the principles involved is necessary rather than knowledge of detailed techniques. The object of this chapter is to deal with these principles; the techniques included will seldom be applicable to many species exactly but the ingenuity of the investigator will enable him to make modifications to suit his resources and his species.

BRINGING HOME LIVE MATERIAL

In order to rear insects it is necessary first to obtain living material from the field. The techniques used in collecting this are the same as those used for material destined for the killing jar but the specimens must be treated with greater care. Insects are usually sensitive to changes in conditions and moving an insect subjects it to changed conditions. The conditions at the surface of a leaf on which an insect is standing in the shade are very different from those it will experience when it flies off out into the sunlit air and those in a glass vial are even more different. Most insects become less active in the dark (if they are normally active in the day time) and if they are cooled. Bottles and vials are particularly prone to overheating in the sun and most insects will die in a few minutes if exposed to the sun in a bottle. Tins are also liable to become hot; cardboard boxes are less prone to overheating but nevertheless they should not be exposed to heat. It is a good rule to use darkened transporting containers whenever possible. Ventilation should not be forgotten, especially for large specimens but it is more important to bear in mind the size of the insect in relation to its container than the actual size of the insect. A large caterpillar in a small bottle will suffer but a hundred tiny parasitic wasps in the same bottle will remain quite healthy.

Live insects should be handled gently. If they are taken up in an aspirator they should not be sucked so that they are pounded against the sides of the intake tube on their journey to the bottle. Do not squeeze caterpillars (or even the hardest beetles) when they are picked up, as, although they may appear to suffer no ill effects, they may die some time later from internal damage.

Make sure that all transporting containers are clean.

Glass or plastic vials are useful; sizes can be chosen to suit the size of insects to be carried. The stopper, if plastic, should have several ventilation holes punched through it. Many small holes are preferable to a few large ones. The stopper can be replaced by a cork which has had a central hole bored right through it. The end of the cork which goes into the bottle can be covered with gauze or cloth. A piece of cheesecloth or other open-mesh cloth can be placed over the mouth of the tube and secured by a rubber band to replace a stopper.

For larger specimens, plastic or glass jars can be used and covered in the same way as the vials. Ice cream and margarine containers are useful. Plastic boxes, obtainable in a great range of size and shapes, such as are used in refrigerators can have holes bored in the lid for ventilation and the lid secured by a rubber band.

Predacious species should be carried singly as they may attack each other if confined together.

Sticks or pieces of bark (or corrugated cardboard) attached to the insides of the box will provide footholds; having loose articles in the box should be avoided. Leaf feeding larvae should always be provided with food, especially if they are to be in the box for more than an hour or two; they often continue feeding in the dark of the box. In the case of Hymenoptera a few raisins, split open, can be pinned securely into the box. Care should be taken not to include too much plant material as this may give off an excess of moisture.

Various forms of portable, well insulated boxes, designed for carrying food on picnics are now marketed. These usually have provision for the insertion of ice, dry ice or other cooling agent to maintain a low temperature in the box. These boxes are useful for bringing home live insect material. If the specimens are placed in glass or plastic containers and these are packed into the cool box the low temperature will reduce their activity, usually without adverse effects on the insect. Some species can be left in such a box for several days without ill effects. It must be remembered that an excess of the carbon dioxide given off from dry ice will kill the insects.

The transport of aquatic insects raises special problems owing to the fact that at least the water in which they live is usually moving and hence is becoming aerated and maintaining natural temperatures. A container with a wide mouth would assist in maintaining the water in an aerated condition but unfortunately water splashes very easily from such a vessel in transit and if the container is left in the sun the temperature rises. Many species can be transported in wide-mouthed jars if some material from the bottom of the pond or stream is included as well as plant material. Aquatic insects can often be transported in wet moss or in aquatic vegetation which has been taken from the water and allowed to drain for a few seconds. This is then pushed fairly tightly into jars or in plastic bags and the insects put into this. The insects are thus kept in an atmosphere with very high humidity but little free water. When they reach their destination they can be floated off into water for sorting. Some insects will travel well in plastic bags filled with pond water and some water weed. The bag can be closed by string or a rubber band after a few specimens have been put in. It is preferable to put only a few specimens in each bag. The bags can be placed in a larger container of water, as this will decrease the likelihood of the bags being punctured or the insects being jarred too much in transit.

SOME PRINCIPLES OF INSECT REARING

Rearing insects is not usually difficult. It is best to rear the insects in an environment which is as close as possible to that of the species in nature, with such modification as may be necessary to make observation possible. The so-called micro-climate is more important in this regard than the general climate as it affects us; for example, the conditions existing at a leaf surface or under a stone are very different from those of the open air or surface of the ground so far as temperature, moisture, light and air movements are concerned. In the vast majority of cases it is necessary to bring the specimens to the laboratory and there to attempt to provide suitable cage conditions in which the insect will develop. This does not mean that the entire environment must be duplicated in the cage. Some species have never been reared successfully in captivity; in these cases there is usually some peculiarity in the environment which is not known or which cannot easily be provided indoors. Sometimes, of course, it is possible to confine the insects in some way in their natural surroundings so that they can be adequately observed. Detailed studies are usually carried out on laboratory-reared specimens whilst a check is kept on caged field material or on the wild population. It is sometimes necessary to rear an insect part of the way through its life cycle in the open and then bring it indoors, or vise versa.

Food

It is usually preferable to feed an insect on the food on which it was found (if a plant) even if other species of plant are known to be acceptable. Whether this is done may be determined by the availability of the food and it is quite likely that the insects will have been collected some way from the laboratory and fresh supplies may not be found nearby. Freshness of food is important (even carrion feeders are fussy about the stage of decay of their meat) and it should be remembered that food and parts of food plants often deteriorate quite rapidly when put in containers or into insect cages. It is essential to maintain a constant fresh food supply even if that provided previously has not been completely eaten. In the normal course of events there should always be more food present that the insects will eat before next feeding time. Plant food can often be kept fresh in a 'deep freeze' refrigerator until needed.

It is better to offer small quantities of fresh food often than to provide a large quantity at infrequent intervals. Insects needing liquid foods, e.g. honey, can have it supplied in the same way as water.

Water

In captivity it is wise to supply water unless it is known that the species definitely does not drink. It can be provided on a tight wad of cotton wool or a sponge which is soaked in water and held in a small bowl (a tin lid or watch glass often suits the purpose quite well). A tight roll of cotton wool inserted into a hole in the lid of a covered plastic jar to form a wick which hangs in water within the jar is a useful continuous source of water. A jar, filled with water and inverted over a flat glass dish, such as a petri dish, lined with filter paper or blotting paper, will also provide a continuous source of water. Although it is usually adult insects which require water, many nymphal and larval forms take it when it is available. Damp cotton wool can become mouldy; when this happens it should be replaced or an unpleasant odour might result. It should also be remembered that the introduction of water , into a cage may raise the humidity level; in most well-ventilated cages this is not of great importance but it may be so in small spaces.

Two methods of providing water for live insects; the water container is placed inside the rearing cage.

Temperature

The importance of heat in insect life cannot be overstressed; their body temperature is not constant but varies more or less with the environmental temperature. In general insects feed, grow and develop more rapidly at a higher temperature than at a lower one. In practice, there is a certain temperature, the optimum temperature, at which growth is most rapid and above or below which the process is slowed down. The optimum temperature is not the same in each species. A few insects have a narrow range in which they can survive but most are fairly tolerant and can be reared at the ordinary prevailing temperatures of the area in which they normally live. It is important to take note of the temperatures at which insects are reared as this could be important in determining the rate of development and length of life and is important information for anyone wanting to repeat the rearing.

Humidity

The amount of water vapour in the air may have important effects on living insects. For example, in some species the incidence of virus diseases seems to be greater at constantly higher humidities. Some species cannot survive if the air is not humid; this applies particularly to soil insects.

Other environmental factors

It sometimes happens that at some particular point, after successful rearing for part of the life cycle, the insects die. In most cases this can be traced either to disease or to some required condition which is not supplied. It should be remembered that insects may vary in their requirements at different times of their lives. The adults may be free flying, the larvae may feed on leaves at the tops of tall trees whilst the pupae may be found in the soil. For each stage appropriate conditions must be provided or, at least, some attempt made to substitute conditions which will allow survival. Food and water have been mentioned and food will be treated a little more fully later. The most common causes of apparently inexplicable failure are often connected with egg-laying (oviposition), copulation, pupation, hibernation and disease.

Some species require rough, others smooth surfaces for oviposition. Some require wet, others dry surfaces and some require quite deep crevices. Some like to be able to creep away completely before laying. Most plant-feeding species require plant material of one or a number of species of plants before they will lay and if the host plant is known, but success is still not obtained, various parts of the plant should be provided. In the normal course of events, of course, little difficulty should be encountered if the natural food of the larva is available to the adult for oviposition. Soil dwellers may require soil of a certain type or degree of humidity, study of the natural habitat will give a lead here. Predators will usually lay if given the conditions of temperature, light and humidity of the natural environment.

Copulation is normally easily achieved, bringing the two sexes into proximity with one another is usually sufficient. Occasionally, however, copulation occurs only when there are several specimens together. Some species need sunlight before they will copulate; others require shade. It is often useful, if a female only is available, to expose her in the open in a small open-mesh cage. Males will sometimes be attracted with remarkable rapidity and a pair can then be allowed to copulate. Some species copulate in flight and so need a big space at this time.

Suitable conditions for pupation should be supplied. Soil is often required and many larvae have a period of wandering after they have completed feeding but before they settle down to pupate. Here again, various surfaces, various degrees of moisture and perhaps a little refuge in which to hide should be tried in difficult cases.

Some species need a period of hibernation at certain times of the year and the most usual requirement for successful hibernation is a dark crevice into which to creep. Rolled up paper, wood shavings or other clean litter, corrugated cardboard and similar materials will often suffice.

Cleanliness of rearing cages is extremely important. Diseases are due to fungi, bacteria and viruses. When the insects die of fungus disease the carcasses may become hard and mummified, spore-forming bodies of the fungus may grow out from the bodies of the insects or the developing fungal tissues emerge and seal the body onto whatever object it may have been standing on. Virus diseases often make the contents of the body fluid and foul-smelling and the cuticle may burst and release the fluid which then dries up. The dried skin may remain for some time. Virus diseases are frequently introduced to the cultures with the food medium. Further infections can be caused by the remains of infected individuals being taken up by other members of the culture. Infected cages can be cleaned with hot soapy water or disinfectant and water, or detergent and water. They can also be washed with a solution of formalin and then washed again with soap and water, after which they should be allowed to stand in sunshine until completely dry.

REARING EQUIPMENT

An insect is usually found near its food. As rearing work always involves the provision of food we shall first discuss some of the techniques which will be found generally useful but modifications will probably occur to the investigator which are more suited to his immediate requirements. Some insect cages have been designed specifically for outdoor work, others for indoor work and some can be used in both situations.

When attempting to rear insects out of doors it must be remembered that as soon as a cage of any sort is placed over a plant, part of a plant, or on the ground, the environment within the cage becomes different from that of the surroundings. The main cause of the change is the decrease in air flow; this results in a rise in temperature in sunshine and in a rise in humidity. If the cage is not sufficiently well ventilated condensation of water will probably take place. Cages should be carefully designed and constructed. The same remarks apply to indoor cages.

Rearing insects associated with foliage

Insects which feed by chewing leaves are probably the easiest to rear. As long as they are kept in well-ventilated cages with an ample food supply and their droppings are not allowed to accumulate they should remain quite healthy and undergo normal growth. They are amongst the insects which can be reared out of doors; they can often be caged on their food plants.

Most outdoor cages in general use consist of fabric or mesh made of various materials, supported by a framework of wire or wood with a means of opening the cage without removal from its position. Materials suitable for cages include metal screening, preferably copper (which is quite expensive) or galvanised wire; ordinary metal gauzes tend to rust very quickly. Recently a variety of plastic materials in meshes of a wide range of sizes have come onto the market which give known amounts of shade. Some of these are very durable. Fibreglass mesh is very useful; lightweight open-mesh cloths tend to rot quickly when used outside. The smaller the mesh, the more difficult it is to see into the cage, but mesh size will be determined by insect size. Whole plants can be covered by a cage or just a small part. Even as little as a single leaf can be enclosed. It is wise to arrange things so that the cage is not supported by the plant alone and that it is supported well enough not to be blown away by wind, remembering always that there is considerable wind resistance offered even by fairly wide-meshed material.

Wire framed gauze cage covering plant on which insects are feeding.

If the cage is to cover a sizable plant such as a shrub or small tree, it is preferable to have the frame of wood or thick galvanised wire and a large door should be included in the design. Such a cage may not need support if it stands on the ground but if it is made with a cloth bottom which can be tied around the trunk of a tree, it can be lifted clear of the ground by a wooden 'gallows' attached to it and planted in the ground. Individual branches can be enclosed in cylindrical cages. In general these can be lighter and made of a galvanised wire framework covered with fabric or mesh; entrance can be gained through one end. The fabric can be extended at either end beyond the cylinder framework and drawn together by a drawstring to close the cage. Many modifications of this type of 'sleeve cage' have been devised, ranging in size from those large enough to cover tree branches down to small, single-leaf cages. Whether support is needed or the cage be allowed to hang on the plant will depend on the size of cage and strength of plant. The simplest type of sleeve cage consists of a fabric bag tied around the end of a branch, this can be used quite successfully with many species.

Sleeve cage over living branch. The cage can be made of any size suitable for the insects being reared.

Leaf-chewing insects reared indoors can simply be placed in a cardboard box with some fresh leaves and the box covered securely with organdie, mosquito netting or cheesecloth. It is better to provide a cage which allows the insects to be seen and which can be cleaned without undue disturbance of the specimens. A simple glass cage can be made from a large jar with a covering of fabric. It is better to use a lamp glass, on a tray or in a flat glass dish or petri dish. The lamp glass is closed at the top with cloth and twigs of the food plant being placed inside in water in a small vase or simply placed in the glass. This arrangement makes cleaning easy. The twigs of food plant should be held in place in the vase by a cotton wool plug which will also prevent the insects from wandering down or falling into the water, which they are prone to do. For species which need soil in which to pupate the lower container can be filled with a sufficient depth of sand or soil and the lamp glass and vase placed on this.

Lamp glass and cover over a small dish for rearing insects in the laboratory.

If the insects need more space they can be kept in a wooden-framed, gauze-covered cage with a vase of food plant inside. As a rule it is unwise to give insects more space in a cage than they need as many leave the food when not actually feeding and frequently they cannot find their way back to it. They tend to accumulate on the brightest side and top of the cage and wander over these until they die.

94

Cages such as the above can accommodate from several to hundreds of specimens, depending on size, but some specimens of each species studied have to be reared individually if accurate information is to be obtained on them. Leaf-feeders can be reared in vials closed by means of a gauze cover or a tight cotton-wool wad. The size of vial depends on size of insect, remembering that in addition to the insect and the food supply a reasonable amount of air space should be available inside the vial. Jam jars and similar receptacles will suffice in lieu of larger vials. If an individual does not eat a great deal of leaf tissue during its lifetime it can be confined on a potted plant covered by a lamp glass or a series of potted plants can be available for each specimen. This arrangement ensures that fresh food is always available, although a small specimen may not easily be found in a cage which is relatively so much larger than itself. A potted plant can also be enclosed in a cylindrical wire-screen cage or wooden-framed cage which can be made to have a greater capacity and better ventilation than a lamp glass or similar covering.

To rear insects which live within the leaf, the leaf miners, a living plant is essential. The adults can be confined with a potted host plant and may be induced to lay eggs. If a leaf containing a leaf miner is brought in from the field the insect can often be reared simply by placing the leaf in a vial tightly closed with cotton wool. Very often the miners, which are usually quite small, can complete their feeding before the leaf dries out.

Most of the species which suck sap (Hemiptera) require a living plant on which to feed. They can be reared in numbers by confinement in a sleeve cage or in a cage over a potted plant in the same way in which leaf chewers are confined. Individuals, especially of the larger species, can be reared in cages over potted plants or seedlings.

A single leaf of a living plant can be housed in a miniature sleeve cage. The body of the cage is made of a cylinder of celluloid, one end of which is attached to a sleeve of cloth while the other is closed by similar material. The leaf is inserted into the cage and the sleeve tied securely around the stem. A similar small sleeve can be made to enclose a twig bearing a few leaves or an ordinary, but miniature, sleeve with two open ends will serve to provide twig-suckers with a length of twig. Many Homoptera embed their eggs in the stem, leaf or other tissues of plants and these should be available within the sleeve. A slightly larger sleeve can be placed around a branch of a small potted plant if the weight of the sleeve can be supported by a wire or wooden cradle. For minute leaf-suckers, such as some of the young Hemiptera and for thrips, cages enclosing only part of a leaf can be used. A simple form of such a cage consists of two pieces of thin wood, one of which has a circular hole cut in it. The complete piece of wood is put on one side of the leaf and the piece with the hole on the other. Between the piece with the hole and the leaf surface is placed a piece of cloth with a hole cut in it to correspond with the hole in the wood. When in use a long rectangular cover slip is placed over the hole; thus, a cell is formed by the circular hole in the wood in which the insect is kept. The two layers of wood, the leaf, the cloth and the cover slip are then held in position by a paper-clip, a rubber band or by a binding of strong cotton. Observation of the insect can be made through the cover slip. A small observation cell can be made from a short piece of wide glass tubing, covered at one end with cloth, and attached to a spring clip which is also connected to a padded disc. The glass tubing forms the cell and is placed against one surface of the leaf whilst the padded disc lies on the opposite side of the leaf; the leaf is thus gripped between the two.

Rearing insects associated with fruit

Rearing insects in fruit is sometimes relatively easy. It is usually the larval stages which occur within the fruit; the adults, usually being free-living, have to be catered for separately and provided with appropriate fruit when they are due to lay eggs. Fruit fly and other pests of fruit, being economically important, have been much studied and techniques have been described in detail in the entomological literature for rearing them on a large scale. To rear other species, such as the less well known insects associated with wild fruit, it will probably be necessary to provide the adult with various parts of the appropriate food plant, such as pieces of stem, leaves, twigs, flowers and fruit in various stages of development if its egg laying habits are not known. This is best done initially on a small scale in a fairly small container indoors so that it will be easier to find eggs or watch the process of oviposition. Some fruit insects start their development on young fruit, soon after flowering; in such cases it may be necessary to confine the adults on a plant with such fruits so that oviposition and development can go ahead on the living plant; others will only attack fruit towards ripening time. Whereas most edible fruits are juicy and will remain fresh for some time after picking, many of the wild fruits have only a little flesh and dry out comparatively quickly once they have been removed from the tree. In these cases plucking a branch and keeping its end in water will help to maintain succulence. It is necessary to make provision for pupation with most fruit insects. If they are reared indoors the fruit should be placed on slightly damp sand or soil, if the fruit is on the tree it will be necessary to remove it to the laboratory when the larvae are ready to leave the fruit. Pupae can be obtained by sleeving the fruit in a hanging sleeve closed at the lower end to which is attached a container of soil or sand.

In the case of hemimetabolous insects which feed on the surface of fruit, they can often be confined in the container with picked fruits and they will remain on the fruit to feed. If the fruit is small several can be put in one container so that if the insects wander they are likely to end up on another fruit. If the fruit is large and the insects have a tendency to wander from individual fruit they can be enclosed in a small celluloid cell, closed by a cover slip and sealed to the fruit by any suitable glue. Individuals can be reared in such small cells.

When fruit falls to the ground and begins to dry out and decay it is invaded by a fauna of its own. This includes not only insects which feed on the decaying fruit but also a range of predator species which feed on them. These insects can be reared by offering them fruit in various stages of decomposition. Many require fruit to be kept moist; others will only survive on fruit which has dried out to some extent. Some of the insects of dried fruits occur as pests in commercial products.

Rearing insects associated with bark

Many of the insects found on bark do not feed on it but are either passing up or down the tree (such as ants on their way to the foliage), lying in wait for prey or resting there. Many of the insects which do live on the bark usually feed on the lichens, algae or fungi which are growing there. These browsing insects can be maintained in the laboratory on pieces of bark carrying the relevant food. They can be kept in glass containers of almost any kind; vials stoppered with cotton wool are useful for smaller species, covered flat glass dishes are suitable when larger pieces of bark are to be accommodated. When feeding is to continue for some time it is useful to cut pieces of bark of appropriate size and, when these have had all the available food removed from them, secure them to the trunks of trees in a situation similar to that from which they came. This will often encourage regrowth of the food medium and the pieces can be used again. This arrangement will often make it unnecessary to seek new sources of food.

Rearing root-feeders

Root-feeders normally spend most of their time in the soil, in the dark. Rearing techniques for soil insects have to include some arrangement whereby light is excluded; observation of the specimens can usually be for short periods only as most root-feeders become agitated on exposure to light and may be adversely affected. Most root-feeders in which the general entomologist will be interested are larvae, especially larvae of beetles. If it is desired to observe these insects in a more or less natural situation they can be enclosed in a soil-filled wooden box which is narrow but deep with a sloping front made of glass, in which appropriate plants are growing. The roots of the plants will grow down into the soil until they make contact with the glass, after which they will grow down along the inclined glass. When the insects are not under observation the front of the box can be covered with thick black cloth. If even a small amount of light is allowed to reach the roots they will grow into the soil away from the glass. The box should not be filled with soil to the top in case the insects go to the surface of the soil and fall from the box. An internally projecting flange of metal around the top of the box will prevent most root-feeding larvae from escaping. The bottom of the box should be provided with drainage holes so that the plants can be watered without danger of the soil becoming waterlogged. Some root-feeders require specific moisture levels for survival so the watering should be adjusted to suit the insects.

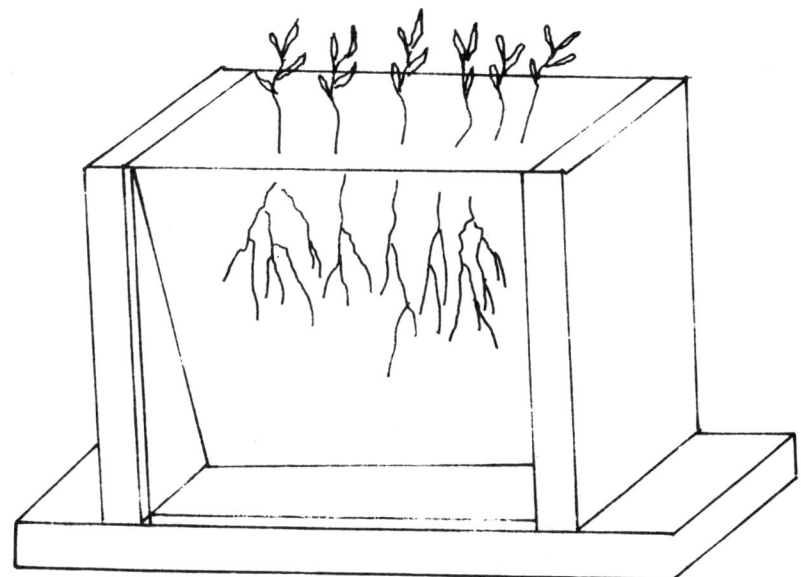

A simple wooden rearing cage with sloping glass front for rearing soil insects. The front should be covered to keep the cage dark when the insects are not being observed.

If it is found that the insects will survive on roots without soil and if a supply of suitable root is available they can be reared quite simply under a lamp glass as in the case of leaf-eaters. The lamp glass and its cover should be kept in the dark.

Root-feeders often thrive on the roots of seedlings. These are easily grown, either by sowing seed in ordinary horticultural seed boxes or by germinating such plants as beans between layers of damp blotting paper or in damp cotton wool. The last two have the advantage that the roots are quite clean; the first method provides larger roots if necessary as the seedlings can be allowed to continue growing almost indefinitely.

For rearing soil insects individually a shallow wooden tray (about 5 cm in depth) can be divided into a large number of square cells by partitions about 3.5 cm apart. Each cell can be filled to a depth of about 2.5 cm with soil, each cell accommodating one insect. The insects are fed on fresh supplies of root, collected from the field and washed or grown specially for the purpose. This method is particularly suited to the rearing of white grubs (larvae of scarabaeid beetles).

In order to obtain eggs from root-feeders it is usually enough to confine the adults over soil, preferably with some growing plants in case they have a preference for laying near roots. An earthenware plant pot is often suitable if the insects oviposit deep in the soil, otherwise shallower containers should be provided. The amount of soil provided should be kept to the minimum acceptable to the insects. The soil should be kept moist unless it is known that the insect prefers dry soil.

Some species simply lay their eggs on the soil surface. In such cases it may not be necessary to provide the adults with soil at all; they can be held in a large lamp glass over damp filter paper or blotting paper on which the eggs will be quite visible when dropped. On hatching, the larvae can be transferred to soil for rearing.

Rearing insects associated with fungus

Many of the insects living in damp environments are fungus-feeders. Psocids feed on bark and leaf fungi, a wealth of forms feed on the fungi of leaf litter and many feed on the fungi which develop in decaying wood and between wood and bark. These forms can usually be reared by giving them conditions in which the fungus will thrive, which is not at all easy in most cases unless equipment is available for the close control of temperature and humidity. The species which feed on the fruiting bodies of the larger fungi, such as mushrooms and bracket fungi, are more easily handled as they can often be reared by providing the adults with sections of the mushroom or bracket fungi on which to lay. Individually they can be reared on pieces of the host fungus in small vials.

Rearing dung and carrion insects

Rearing insects associated with carrion and dung is not difficult. It is important in the case of dung that the dung be from the correct group of animals and at the correct stage of decomposition after dropping as dung usually has a sequence of species which feed on it. This also applies to carrion-feeders but the actual species of corpse is not as important as the specific origin of the dung.

Flat glass containers are useful for rearing dung insects in batches and vials for individuals. Most dung and carrion feeding insects need soil in which to pupate.

Rearing aquatic insects

The aquatic environment is very diverse and as a result there are a great many different techniques for rearing water insects. Some of them are not easily reared in captivity, others are very easily reared in ordinary glass aquaria.

The insects living on the surface of the water can be reared very easily indoors in covered glass dishes, filled to about half way with water, if they are provided with freshly killed insects on which to feed. For egg laying, water plant stems and leaves can be provided. Carnivorous species living below the surface must usually be provided with living food; the most easily obtained of such are mayfly nymphs, mosquito larvae or water fleas (*Daphnia*). It is possible to feed some of the aquatic insects on maggots, mealworms, pieces of earthworm or the various live foods used by aquarists for feeding their fish e.g. *Tubifex*. These are obtainable from aquarist's suppliers. Insects which live in the mud at the bottom of ponds and streams should be provided with this when in captivity as well as quantity of water weed amongst which they can crawl. Insects normally living in hidden situations in the water become overactive if confined in a container which has nowhere for them to creep away. Many of the aquatic insects are weed feeders and these, of course, must be provided with appropriate weed species; aquatic plants will remain in suitable condition for some time, there being no wilting underwater. In general terms, if the natural environment can be reproduced in miniature so far as vegetation, sand, mud or stones is concerned aquatic insects can be reared. Most species will tolerate tap water if the habitat is arranged correctly and if the water is kept sufficiently well aerated. It should be allowed to stand in a bucket for two or three days before use. Some species, however, will only thrive if the water is stagnant, slow flowing or fast flowing, according to that of their natural habitat. Stagnant water is not difficult to provide, especially if the natural habitat is used as the source of bottom material, vegetation, water and specimens. Many slow-water species will survive in still water if it is well aerated. If moving water is required the insects can be reared in a metal tray placed in a deeper receptacle which has an overflow outlet a little above the level of the wall of the metal tray. Water is allowed to pour into the outer container, flow around and into the tray and find its way out by the overflow which should be gauze-covered to prevent the escape of specimens. Small pumps are obtainable from aquarist's suppliers which will circulate the water in an aquarium quite rapidly; these are suitable for species in which the rate of water movement is not high. As many nymphs or larvae have to leave the water prior to emergence of the adult or for pupation, objects should be provided which protrude above the surface of the water in rearing aquaria so that they can creep out. Aquaria should be covered at all times, preferably with a fine mesh screen as most adults of aquatic insects will take flight. This is especially so if the conditions in the aquarium are not ideal for the insect.

Aquatic insects are frequently reared in field cages. These take care of many of the minor troubles likely to be encountered in laboratory rearing. The simplest cage for field use is the so called 'pillow cage'. This consists simply of a more or less pillow-shaped cage made of metal mesh (non-rusting materials should be used). This is secured in the stream with the lower part in the water and the upper part above the surface. It can be staked in position in stream or pond. The insects are placed inside this and fed appropriately. Perhaps a more suitable type of cage, which can be made in a large size to hold several specimens or in a small size for individuals, consists of a cylinder of copper gauze with one end of the cylinder closed by a metal or wooden base. The other end is closed by means of the gauze being drawn together and tied. This cage is likewise partly submerged. Stones or plants can be placed in the cage; the mature insects crawl up the gauze above water level when emerging, and can be removed from above by untying the gauze.

If it is merely desired to obtain adults from mature nymphs or larvae these can be brought to the laboratory and allowed to emerge in small aquaria. In all aquatic insect rearing it is far preferable to use water from the natural habitat of the insects as this will contain mineral constituents in the proportions to which the insects are accustomed.

Rearing insect parasites

The rearing of the parasites attacking insects can be one of the most fascinating activities. In practice it is really the rearing of two insects, as, in order to study the development of the parasite, it is necessary to maintain the host as well. We shall deal here with the rearing of the internal insect parasites of insects (sometimes referred to as the parasitoids).

In rearing parasites it is necessary first to have at least some knowledge of the life cycle and annual cycle of the host as specimens of the appropriate stages of the host, known to be parasite free, must be available when required. It is usually necessary to carry out a preliminary study of the life history of the host as it seldom happens that a detailed description of this is available in the literature unless the host happens to be a well-known species of some economic or academic importance. The first step in parasite rearing is to rear the appropriate host and to obtain an idea of when they are available in the field; most insects fluctuate in their abundance and it is usually necessary to rear the hosts used in parasite work in order to be sure that they are parasite free; host species captured in the field may have been attacked already, either by the species that it is intended to rear or by some other species of parasite. On the basis of the knowledge obtained in the rearing of the host alone it should be possible to arrange for host material of a suitable age for parasite attack to be available at the time when the parasite adults are active. Parasite material is usually obtained by bringing in quantities of specimens of the host from the field and rearing these; some of them will probably yield parasites and these form the basis of the parasite stock to be used for further rearing and study.

As a general rule the parasite life history and annual cycle of activity is very well synchronised with the parasites active at a time when the appropriate stage of the host species is available. It is as well to be on the lookout for this correlation of cycle between host and parasite so that conditions suitable for both can be provided at all stages of the life cycle and at all times of the year. Quite often it will be found necessary to rear the parasites for one season in order to obtain some of the basic information concerning their life history, following this up in subsequent seasons with more detailed work which can only be carried out when the elements of the life cycle are known.

The simplest parasites to rear are those which complete their development within an insect egg. For rearing these a continuous supply of parasite-free host eggs is needed. Most egg parasites will attack the eggs if enclosed with them in a glass vial; many species require the eggs to be fairly recently laid but some will accept advanced eggs. Most egg parasites have a very short life cycle and several generations can be reared during the period for which the host is normally available in the field. After this there may be a long period of inactivity on the part of the parasite; this long period of waiting for more hosts may be passed as a larva or a pupa within a host egg shell or may be passed as a hibernating adult.

In order to obtain a sufficient supply of host eggs a culture of the hosts should be kept going in a parasite-proof cage. As adult hosts lay their eggs these can be removed for exposure to the parasites. As the parasites develop within the host egg it will be necessary to open up eggs at known times after attack by the parasites in order to observe the various stages of the parasite. In order to obtain an initial stock of egg parasites, eggs which have been laid in the host culture can be placed out of doors for a day or two in a situation similar to that of the normal egg-laying site of the host. The eggs are then brought indoors again and allowed to complete their development. A proportion of the eggs or egg batches so treated may yield egg parasites instead of the young of the host, indicating that the eggs were attacked whilst exposed out of doors.

Insects which attack larvae may lay eggs in the earlier instars of the host, thus allowing the parasite to complete its development before the host pupates. Some parasites seek out and attack the pupal stage. If the most suitable stage for attack is not known several hosts in various stages of development should be exposed to the attack of the parasite.

It is usually sufficient, to ensure attack, to introduce mature parasites into a cage containing the host specimens. Apart from the very small species of parasites, ample room should be allowed in the cages in which the attack is to be made to allow the parasite room to fly around quite freely. If the host is a plant feeder it should be caged with some of the plant, not only to provide food for the host but it may be necessary to assist the parasite in finding the host. This is because some parasites first seek out the precise environment of the host and, having found this, then set about searching for the host insect. Also, many parasites require room for mating. Unless a parasite is seen attacking a host there is seldom any way of telling which of several specimens exposed to attack have been parasitised. In some Tachinid flies the eggs are placed externally on the host where they can be seen as small white objects adhering to the host cuticle.

The most convenient cages in which to arrange the attack of parasite on host are the cages of the type in which the host is normally reared. Some modification may have to be made to prevent escape of parasites if the parasites are much smaller than the host. Some parasites will only mate and actively seek and attack their hosts in sunshine. Food should be provided for the adult parasites as well as the hosts; in the case of hymenopterous parasites honey and water or raisins are usually satisfactory; in the case of parasitic flies sugar water or honey may be adequate. Lump sugar and water may suffice for some species. Lump sugar has the advantage that it is provided dry and there is less chance of the insects becoming sticky. Some flies and wasps become sticky when provided with honey or sugar water and in their efforts to clean themselves spread these substances all over their bodies, eventually incapacitating themselves.

Parasites which attack the pupa will normally do so if placed in a container with one or more living host pupae.

Having arranged the attack of the parasite on the host, the host is reared in the same way as it would have been had it not been parasitised. Most hosts do not behave abnormally when parasitised apart from taking more food than usual as the parasite develops. Most parasites complete their larval development within the host body leaving it just prior to pupation. They may emerge from the host and spin a cocoon beside the remains of the host body; they may move some way from it before doing so or they may drop to the ground and enter the soil or vegetable litter on the surface of the ground to pupate. It is, of course, necessary to provide appropriate places for pupation within the rearing cages. If these are not known a little fine soil or sand, with a little crumpled paper or corrugated cardboard on top of it, can be provided as substitutes for natural pupation sites.

Although some parasites which attack the larvae of insects live on the outsides of their hosts and their development may be followed quite easily, most insect parasites are internal and, as in the case of egg parasites, it is necessary to kill and dissect a series of hosts at known periods after attack. It is not, of course, necessary to observe an attack to establish the date of attack. If a number of unparasitised laboratory-reared hosts is exposed to attack for about twenty-four hours and then removed to other rearing cages without parasites, they can be killed and dissected at intervals as a series to observe the sequence of development of the internal parasites, some being allowed to continue development until the parasite has completed its development.

Rearing those parasites which attack the host in its pupal stage is fairly easy as long as a supply of host pupae can be maintained. The principles and methods involved are similar to those used for rearing egg parasites, allowance being made for the difference in size and for the fact that many parasites leave the host pupa to pupate themselves whereas most egg parasites pass their pupal stage in the host egg.

Rearing predacious insects

In every habitat, whether terrestrial or aquatic, there is a large number of species which are predacious. These predators form a very important part of the community of animals in nature and, with the parasites, play a large part in keeping down the numbers of other species. To rear predacious insects it is essential to have a ready supply of living food, which, in most cases, consists of living insects. With most species rearing is easy if a food supply is easily available. Some predacious insects will feed only on a single species of prey, or on a few closely related species; others feed on a wide variety of species in a given habitat. Some species which are normally predacious can be reared on meat in the laboratory. Some species will take their prey only if it is moving; others will only take it if it is flying but in most cases it is enough if the prey be alive.

Many predacious adults inhabit a different environment from their young and their food requirements may differ considerably. It is also necessary to take account of the requirements of the females where oviposition is concerned and although many predatory species will deposit their eggs almost anywhere some require special conditions.

Eggs of damselflies can be obtained by noting where the females lay; they do this in plant tissues under the water surface and they can be seen dipping their abdomens below the water whilst clinging to a plant above the water level. Some species will lay if the tip of the abdomen is held in water. A wooden-framed gauze-covered cage will serve well enough for many groups such as the scorpionflies, lacewings and some of the predacious flies and beetles. In small cages it is easy to provide enough food by emptying the contents of a sweep net into the cage. Observation will then give an idea of which insects are preferred and these can be collected. It may be necessary to provide soil, sand or, in some cases, water or mud in the bottom of the cage in which eggs are to be laid. This should preferably be in a removable container, such as a shallow dish, so that the eggs can be found and examined without too much searching. Some predators will place their eggs on damp filter paper. Some require, or prefer, vegetation on which to oviposit. If requirements are not known it is a good idea to provide the conditions known to be required by related species or other species of the same family living in similar habitats. In some cases it will be necessary to provide a variety of oviposition sites and to adopt the one chosen by the insect, for use in future rearings.

Species which feed in flight or which snatch insects from vegetation very often feed on adult insects and these can be supplied at intervals without worrying about providing a food supply for the prey.

Where the prey consists of larvae, or of insects which cannot be separated for long from their own food source, it may be necessary to maintain a culture of the food species. Later in this chapter some suggestions are given on how this may be done for some generally useful prey species. Failing a ready supply prey will have to be collected from time to time. For a species which feeds on, say, caterpillars, beating and sweeping may provide a sufficient supply; for aphid feeders, broad beans, roses and cabbages often carry good populations of aphids at the time of year when their predators are most abundant and the field population of these prey insects can be culled to feed the predators. Whilst a predator should not be without food for any length of time, it is not necessary to supply a great excess of food as most predators are fairly active and will find their food in the cage. In the case of moderately large to small species, a great many of which are quite amenable to being reared individually in glass vials, a few prey are all that is needed, these being replaced whenever the supply has been depleted. Predators frequently become cannibalistic when confined with one another in small spaces, even when ample food is available. It must be remembered that predatory species usually have quite powerful mandibles and some may eat their way out of a container which is not adequately stoppered. This applies especially to some of the larger predacious beetles.

Some suitable prey species
Apart from collecting a miscellaneous lot of insects and offering these to the species being reared, it is sometimes possible to collect specific species in sufficient numbers to maintain the laboratory colony. Ants are taken by some insects when other food is not available although ants are fast moving and many species appear to be left alone even by starving predators. Ants can be taken up by aspirator as they move along their runs between the nest and a food supply. If suitable food (e.g. jam or honey) is placed somewhere near an ant colony it is usually not very long before a column of ants is established and this forms a source of supply which can be culled. By locating nests of those species which build under stones it is possible to obtain ant larvae and pupae (the 'ant's eggs' of the aquarist) and these form suitable food for some predators. If an ant nest under a stone is not exposed for more than a minute or two whilst some larvae and pupae are removed by aspirator and the stone is replaced carefully and quickly in exactly the same position the same nest can be visited periodically. It is best not to visit one nest too often as this may cause the ants to vacate the site. From time to time certain insect species become abundant. For example, caterpillars may occur on a particular species of bush or tree in large numbers; such outbreaks can sometimes be taken advantage of and the specimens used for food. This often comes as a welcome change of diet for species which are being reared on laboratory-cultured prey.

Woodlice are often common in damp situations and if a few bricks and pieces of wood are piled up in a damp spot the congregations of woodlice will serve as a good source of food for such insects as ground beetles.

Most of the requirements of predacious species can be met by providing specimens of appropriate size from laboratory-reared cultures, and these can be used as a matter of routine although it is also advantageous to use other sources of supply, not only in order to vary the diet of the predator but also to conserve stocks of the laboratory-reared material for use during those periods when other food is not available, such as may be the case in periods of cold or rainy weather.

Cockroaches are suitable food for many species and if a large species, such as the American cockroach (*Periplaneta americana*) is reared the various stages of the insect will supply quite a wide size range of food. The smaller German cockroach (*Blatella germanica*) is also a useful species. Cockroaches are easily reared by enclosing some adults in a container with a good supply of food. They will eat almost anything but they thrive on a medium made up of 50 per cent ground whole wheat, 45 per cent dried skim milk and 5 per cent baker's yeast. A quantity of this can be provided in a container such as an aquarium covered with cloth or gauze or in a large glass jar. Cockroaches readily climb glass and chew through cloth and a band of petroleum jelly around the top below the rim of the container will help prevent them reaching the top and making an escape. Water should be provided. The cockroach culture should be kept in a warm and dark place. The container must be cleaned from time to time. This is made easier if the food is placed in an accessible container such as a tin or box. Crumpled newspaper should be placed in the container amongst which the insects can hide when not active to provide a place in which the females can deposit their egg-capsules.

Larvae of the wax moth (*Galleria mellonella*) are easily reared and form a useful food for some insects. Old discarded brood comb from bee hives can be used for food. The eggs are laid on the old comb and the young larvae bore into the wax. It is important to have the wax washed free of honey. This is a prolific species and it may be necessary to reduce the colonies by careful removal of pupae, allowing only a small number of each generation to mature and reproduce. The larvae develop rapidly in warm conditions. Any container can be used, but arrangement should be made for adequate ventilation of the container, particularly as the larvae grow; they tend to wander from the wax and it is essential to make sure that the container is made so as to prevent the escape of small larvae. Milk bottles can be used to house the cultures, the bottles being closed with fine-mesh wire gauze to prevent loss of small larvae and, at a later stage, a slightly coarser mesh to allow ventilation but prevent escape of larger larvae.

The meal moth (*Ephestia kuhniella*) and its larvae also form a useful food supply for caterpillar-eating and general predators. It is easily reared in large quantity. The moth larvae feed on whole wheat flour and an aquarium or wooden box, with a base measuring about 30 cm × 20 cm deep will form an adequate rearing container. A shallow layer of flour is placed in the container with some of the adults. Eggs will be laid and the young larvae will begin feeding on the flour. As they do so, they spin silk which matts the flour in a characteristic manner. The larvae leave the medium to pupate on the sides of the container in a silken cocoon. It is advisable to maintain the humidity at a fairly high level in the container and this can be done by placing a cloth-covered dish or jar of water in the container. Inspection of the culture should be made regularly to make sure mould growth does not develop on the food medium as this tends to reduce the caterpillar population. The presence of mould indicates that the humidity is too high. This species is also very susceptible to attack by hymenopterous parasites, and the presence of adult wasps in the culture boxes indicates that a falling-off of the population in the culture can be expected. In the event of this happening a new culture should be started, taking care to use fresh flour and making sure that the container is covered so as to exclude any insects of the size of the parasite. Initial failure to set up an adequate culture should be followed up by further attempts, using various levels of humidity as this is one of the most usual causes of the failure of cultures to become established. From time to time new flour should be added to that in the container to provide additional food for the developing larvae. It is sometimes possible to maintain a single culture supplying adequate quantities of larvae and moths for some time. When there are signs of a diminution in the culture, a fresh one can be started using adults from the old one. Should disease be suspected in the cultures, fresh cultures should be started with adults from elsewhere. Flour beetles and mites may appear in the cultures with resultant loss of larvae. These may destroy the meal moth eggs but the likelihood of their occurring can be reduced by using fresh, clean flour, using clean containers or, if old containers are to be used, by sterilising them by heat treatment if possible.

A supply of small meal worms (*Tenebrio molitor*) forms a useful food. These larvae have fairly hard integuments. They are easy to rear in metal or glass containers. It is important that the sides of the container be very smooth to prevent their escape. It is preferable not to cover the container. Boxes or aquaria about 60 cm × 30 cm × 30 cm are adequate. The beetles can be fed on chicken mash. This is put into the bottom of the container in a layer about 1 cm deep and covered with a piece of cloth. On this is placed a little more mash and on this another cloth. This should be built up to about six layers. Larvae or adults are introduced into this medium where they should feed and reproduce readily if a little water is sprinkled on the medium each day. The sprinkling can be reduced if pieces of apple, potato or lettuce leaves are placed on top of the medium to maintain a moderate moisture level. When the medium has been eaten and powdered it can be removed and some placed in the bottom of another container with fresh mash and cloth layers on top. The powdery residue will contain beetle eggs from which larvae will hatch and can be used to start a few cultures at intervals after one another so that larvae of various sizes are available most of the time.

Flour beetles (*Tribolium*) can be reared easily and are smaller than meal worms. They can be reared on whole wheat flour in smooth sided dishes or jars. They will thrive under conditions similar to those described above for the meal worms but with whole wheat flour as the food medium.

CHAPTER 6

IDENTIFICATION

A S SOON AS ONE sees a strange object, the question asked is 'What is it?' The same applies to insect specimens. In many cases the mere mention of the name satisfies, but a name is an extremely important thing in entomology. It replaces, in a sense, a description; when we wish to discuss a particular species it is very convenient to have a short name rather than have to give a long description of it each time we refer to it. As we need precision in entomological matters we must be precise in our insect naming, hence the importance of precise scientific names rather than possibly ambiguous common names. The name gives us access to all that is known about the insect. We can look it up in scientific literature and find out what is known about it. It also tells us where the species has been placed in classifications and also which species are related to it. Indirectly, therefore, the name of an insect helps us to find out a lot about it as well as its relatives.

THE IDENTIFICATION PROCESS

Identification, or the correct naming of a specimen, is thus of fundamental importance in entomology, faulty identification can lead to erroneous conclusions. It is important for ecologists to know which species are being dealt with as closely related species may differ in small but important points of behaviour and reactions. Species which are pests must be correctly identified so that previous work on controlling them can be looked up and in parasite studies the names of both the host and the parasite are important.

Identification is easy in the case of large, conspicuous or characteristically marked species. When there are many species which differ from one another only in details it is more difficult. Every entomologist should attempt to identify his own specimens *as far as possible*. The number of species in existence is enormous and no one can hope to carry out identification in all groups. Most people specialise in a small group, say an order, a family or some other natural group of species. No matter what group is being specialised in, everyone should be able to identify his own specimens at least so far as the family level.

For this, a general knowledge of insect anatomy is needed, which can be obtained from one of the standard textbooks on entomology, some of which are referred to in the 'Further Reading' list in this book. Having decided on the group to which a specimen belongs it is then necessary to turn to more specialised entomological literature for further identification. Experience with a group is important and the more practice in identification one has the more likely it is that the identification will be correct. The average entomologist can soon place many specimens directly into a family.

In identification the first step is to decide upon the order to which a specimen belongs. After a little experience this step is usually omitted altogether as the orders can be recognised at once. To the beginner, however, the task of deciding to which order the specimen belongs may be a difficult one. The orders have been listed and broadly characterised in the synopsis of the orders given in Chapter 2, but it would be tedious and inefficient to work through such a synopsis for each specimen. Instead a 'key' to the orders is used. A brief 'key' to the orders of insects will be found at the end of this chapter and many textbooks include similar keys to orders and families.

'KEYS' TO ORDERS OF INSECTS

A key is virtually a table of characters, so arranged that as the user works his way through the key he is constantly presented with two sets of opposing characters, and at each point in the key he has to decide to which of the alternatives his specimen corresponds. A simple example will make this clear. Keys can be arranged so that they lead to the identification of orders, families, genera or species, or indeed, any kind of grouping.

Example: (The example given is that of a key which would enable a user to identify genera within the family Calopsocidae of the order Psocoptera).

1. Fore wing of normal form, about three times as long as wide 2
   Fore wing broadly oval, twice as long as wide................................. 3
2. Veins in fore wing joined by a network of veinlets .... *Neurosema*
   Only vein R and its branches joined by network, other veins normal .... *Dirla*
3. Third segment of antenna thickened and very hairy .... *Mindaus*
   Third segment of antenna not exceptionally thickened and with normal hair arrangements... *Calopsocus*

We shall assume that the user knows that the specimen in front of him belongs to the family Calopsocidae having 'run it out' to this family in a textbook key (which would, of course, be done in the same way as using this key). Starting at the *top of the key* we take the first 'couplet', that is the first pair of contrasting characters. If the specimen has a 'fore wing of normal form, about three times as long as wide' we then pass on to couplet No. 2, because opposite this set of characters is placed the figure 2. Had the specimen had the wings 'broadly oval, only twice as long as wide', we would pass on to couplet No. 3, as instructed opposite that character set. We shall assume that our specimen corresponds with the second half of couplet No. 1. We then, therefore, pass on to couplet No. 3 (ignoring couplet No. 2), and see to which half of this couplet our specimen corresponds. If the third antennal segment is thickened, we assume that our specimen belongs to the genus *Mindaus*, if not, that it belongs to *Calopsocus*. Keys vary in length and many characters may be used. Some keys are more efficient than others, this depends largely on the insects concerned and the knowledge and the skill of the person devising the key. Keys are arranged in various ways but the user is always presented with a choice of characters until he finally arrives at the point where he is given a name instead of another couplet number. The example given above could, for example, be arranged in either of the ways given below, e.g.

1. (4) Fore wing of normal form, about three times as long as wide.... 2
2. (3) Veins of fore wing joined by a network of veinlets ..... *Neurosema*
3. (2) Only vein R and its branches joined by a network; other veins normal .... *Dirla*
4. (1) Fore wing broadly oval, only twice as long as wide................... 5
5. (6) Third segment of antenna thickened and very hairy .... *Mindaus*
6. (5) Third segment of antenna not exceptionally thickened and with normal hair arrangement.. *Calopsocus*

In this arrangement each set of characters is numbered separately and the contrasting pairs separated from one another but are referred to by the number of the opposite half of the pair being placed in brackets.

The key could also be arranged as follows:

A.   Fore wing of normal form, about three times as long as wide.

B   Veins of fore wing joined by a network of veinlets *Neurosema*

B_1 Only vein R and its branches joined by a network; other veins normal.. *Dirla*

AA. Fore wing broadly oval, only twice as long as wide.

BB   Third segment of antenna thickened and very hairy *Mindaus*

BB_1 Third segment of antenna not exceptionally thickened and with normal hair arrangement..................................... *Calopsocus*

This arrangement is not used as much now as in earlier literature; it is only suitable for fairly short keys as the amount of indenting which can be done for each set of couplets is limited by page size.

Other forms of key, such as pictorial keys in which the characters are illustrated, are also sometimes encountered but are neither as common nor as compact as the types given above. For practice a beginner should obtain specimens of known species and run these out in a key; it will then be possible to detect any points in the key at which he is likely to make a mistake.

The next step in identification is to determine the genus and species to which a specimen belongs. This is not usually so easy. If the family has been monographed recently the monograph will probably contain keys, even to species. A monograph is a book (or a paper published in a scientific periodical) which contains a summary of the information available on the particular group with which it deals. There will probably be descriptions, keys, illustrations and information on distribution of the species. It will be appreciated that many groups of insects have *never* been monographed, some were monographed many years ago and so the monograph, although of use as a starting point, may be very much out of date.

Some groups of insects have always been more popular with collectors and it is these, such as butterflies, dragonflies and some families of moths and beetles which are the easiest to identify. Illustrated books on butterflies are most numerous so they are amongst the easiest insects to identify. In order to find out whether there has been a recent (or not so recent) monograph or the literature published on the insects you are interested in it is necessary to consult the *Zoological Record*.

The *Zoological Record* is a publication issued by the Zoological Society of London and one volume appears each year; the first publication was 1864. All natural history libraries should have this publication and as it is usually housed in the reference sections it can usually be consulted by most people but not removed from the library. The 'Zoo Record', as it is usually called, consists of several sections, one of which deals with the insects. It consists of a list of papers and books published during the year to which the volume relates. There is an index to subjects, which refers across to the list of papers so that references to any subject can be traced. Also, there is a list of the families and species mentioned in the literature, again referred back to the list of papers and an index indicating which papers refer to each area of the world. The *Zoological Record* has remarkably good coverage of entomological literature and, although there is a time lag of a few years in publication at present, it is an indispensible work of reference to the serious entomologist. By referring to it, the place, date and author of information on the topic or insects in which you are interested can be traced and the original periodical or book can then be searched out in the libraries.

Entomological literature is now very extensive and in almost every country of the world each year there are published a great many so-called 'papers' in addition to books. Most entomological or zoological societies publish journals in which their members publish papers giving the results of their work. Most museums and research organisations publish their own journals and there are a number of journals which are published by publishing companies. It is clearly impossible to obtain all of these publications personally but most scientific organisations which publish a journal have in their libraries those of other organisations and these can be referred to. Some scientific institutions have large libraries; others have small ones. It is possible for most local libraries to borrow a book or periodical from another one on behalf of an individual. In this way a paper or book traced through the 'Zoo Record' can be seen.

In addition to the *Zoological Record* there has been published annually since 1969 *Entomology Abstracts*. This is a journal which lists papers on insects published each year, much as the *Zoological Record*. Its coverage is not as wide as the *Zoological Record* but it does present a short synopsis of content for each paper listed.

As soon as an entomologist begins to specialise he accumulates literature on his special interest. Some of this will be in the form of books which he has purchased but much of the literature which he will require will be in the form of 'papers'. It will be necessary for him to seek these out in libraries, having obtained the references to them in the 'Zoo Record' or having heard of them from some other source, such as a colleague interested in similar matters. Authors are usually given a few copies of their papers when they are published. These are known as 'reprints' and most authors are glad to present a reprint to someone genuinely interested in it. Politely requesting a copy of a paper will often result in a reprint being given. In this way papers dealing with the subject or group of insects in which you are interested can be acquired. If an author's supply of reprints is exhausted it may be necessary to obtain a photocopy of a paper (this is usually fairly expensive for a long one) if the paper is sufficiently important to warrant needing a copy of your own. A synopsis of less important papers on a subject can be made which will serve as a reminder of the general content of the paper which can be referred to again in full if necessary. There is no objection to making such synopses of papers in libraries providing that they are not subsequently published.

Returning to the process of identification, if a monograph on the appropriate group has not been published, it is necessary to refer to the many original publications on the group, traced through the *Zoological Record*. If a monograph is available which is not very recent this may be used in conjunction with recent papers published which will allow identification of species described after the publication of the monograph. Some groups have been catalogued, that is, a list of the species has been published with references to where the original and subsequent references can be found. Such catalogues are very useful and save time.

Regardless of how the initial identification is carried out it is necessary to check the identification by carefully comparing the material with the original description and with any later authoritative re-descriptions. (This is not *always* necessary, as in some groups e.g. butterflies from areas which have been well worked for a long time the species are so well known that there cannot be confusion; this, however, is true only for very few groups). Having checked the material against the description point for point it is also a good idea, especially for the inexperienced worker, to have a close look at specimens which have been identified as the same species by an established authority or to obtain confirmation by other collectors of experience and knowledge; wherever possible specimens should be checked against identified material in existing collections.

Identification is quite easy in some groups but in less well known groups it may be quite a heavy task which will take much time, effort and reference to scientific literature. In any event, it is a very interesting and satisfying activity and the increase in knowledge with each case is most encouraging. With less well known groups it very often happens that the specimens cannot with certainty be referred to any described species. This raises the possibility that the specimens may belong to an unknown species. The problem then becomes one for a specialist in the group concerned.

Sooner or later, most entomologists decide on a particular group of insects in which to specialise or they decide to specialise in the study of some particular *problem* rather than on a group of insects. In the latter case, after the initial sorting of the specimens collected whilst studying the problem they can be sent to specialists for identification.

Having identified a specimen, it should be labelled with its name and the name of the person who identified it, e.g. *Danaus plexippus* L. det. J. Smith, 1960. The abbreviation 'det.' stands for 'determined by' and is followed by the name of the identifier and the date (year is sufficient) of identification.

KEY TO INSECT ORDERS

The following key to orders includes adult and immature forms but there are limitations which should be borne in mind. Tremendous diversity of form is found amongst insects and a key which would lead infallibly to all forms would be extremely complicated and long. Also, a few aberrant forms will be found which will not run out correctly. In any case the specimen should be checked against the characters given for each order in the synopsis at the end of Chapter 2. It should be noted that in one or two cases where larval forms are concerned the specimen may be referable to one of several orders (e.g. couplet 57). Couplet 34 offers a choice of three sets of characteristics instead of the usual two. There should be no confusion here, however, as the three possibilities are quite distinct from one another. Unfamiliar terms should be checked in a textbook, but most of those used are self explanatory or have been explained earlier in this book.

KEY TO INSECT ORDERS—ADULTS AND IMMATURE FORMS
(adapted from Brues, Melander and Carpenter 1954)

1. Fully developed wings present 2
   Without wings, or with reduced wings or wing buds 33
2. Fore wings of thicker texture, at least partly so, than hind wings... 3
   Fore wings membranous, not thicker than hind wings 12
3. Fore wings with veins recognisable; hind wings not folded transversely as well as longitudinally in repose; mouthparts various................... 4
   Fore wings without recognisable veins; hind wings folded transversely as well as longitudinally in repose; mouthparts always adapted for chewing 11
4. Mouthparts specialised for sucking, in the form of a set of stylets contained in the labial sheath, the proboscis thus formed being jointed 5
   Mouthparts adapted for chewing 6
5. Head with proboscis arising from the head so as to project downwards; fore wings thickened in basal part, membranous in distal part.. HEMIPTERA
   (suborder HETEROPTERA)
   Head with proboscis arising from hind part, so as to project backwards between legs; fore wings of similar texture in both basal and distal

parts .... HEMIPTERA
(suborder HOMOPTERA)

6. Hind wings not folded in repose; thickened basal section of wing very short with a transverse suture; similar to fore wings .... ISOPTERA
Hind wings usually folded in repose, without basal transverse suture; broader than fore wings.. 7

7. Antennae usually thread-like; prothorax large; cerci present; fore wings usually long.:.. 8
Antennae short, one or more segments with a lateral process; cerci absent; fore wings small, hind wings relatively large and membranous... STREPSIPTERA
(males only)

8. Hind femora normal, not larger than those of fore legs; wings held more or less flat over body in repose; no sound-producing organs........... 9
Hind femora much enlarged, larger than those of fore legs, enabling the insect to jump; if not, then fore legs modified for digging; wings usually held roof-wise over abdomen; sound-producing organs usually present... ORTHOPTERA

9. Body elongate; head clearly visible when insect is viewed from above.. 10
Body oval, somewhat flattened; head more or less concealed when insect is viewed from above; legs with large coxae and with spiny tibiae.. BLATTODEA

10. Prothorax larger than mesothorax; fore legs adapted for seizing prey; cerci many segmented .... MANTODEA
Prothorax short, fore legs similar to remaining legs; cerci not segmented.. PHASMIDA

11. Abdomen with forceps-like organ at apex; antennae long and slender; fore wings short, most of abdomen exposed; hind wings almost circular, folded fan-wise.. DERMAPTERA
Abdomen without apical forceps-like organ; antennae of various forms; fore wings usually covering most of abdomen .... COLEOPTERA

12. Four wings present... 13
Two wings only present 31

13. Wings reduced to narrow almost veinless, strap-like organs with a fringe of long hairs, tarsi two segmented or unsegmented, with a swollen tip; mouthparts asymmetrical, adapted for piercing and scraping plant tissue; cerci absent; small species .... THYSANOPTERA
Wings broad, usually with veins; last tarsal segment not swollen at tip.. 14

14. Wings, body and legs clothed with scales; mouthparts reduced, a coiled sucking tube formed by the maxillae; sometimes mouthparts vestigial .... LEPIDOPTERA
Wings, body and legs not clothed with scales, but may have some scales amongst otherwise hairy body covering 15

15. Hind wings with a distinct anal area folded fan-wise in repose; hind wings nearly always longer and broader than fore wings; antennae conspicuous; wing veins usually numerous; nymphs or larvae usually found in water.. 16
Hind wings without a distinct anal area; hind wings not larger than fore wings 18

16. Tarsi 5-segmented, cerci not conspicuous..................................... 17
Tarsi 3-segmented, cerci usually conspicuous; prothorax large
.... PLECOPTERA

17. Wings with several crossveins along front cell; prothorax fairly large.. NEUROPTERA
(part)
Wings without such cross veins; prothorax small... TRICHOPTERA

18. Antennae short and inconspicuous; wings with network of veins formed by main veins and numerous cross veins; nymphs found in water.. 19
Antennae usually large, if small then wings have few cross veins or the mouthparts are modified for sucking; young stages terrestrial or aquatic, but usually terrestrial 20

19. Hind wings smaller than fore wings; abdomen ending in filaments; tarsi usually 4- or 5-segmented .... EPHEMEROPTERA
Hind wings similar to fore wings; no filamentous processes at end of abdomen; tarsi 3-segmented.. ODONATA

20. Head prolonged below into a beak with mouthparts adapted for chewing at the apex of the beak; (hind wings not folded in repose) .... MECOPTERA
Head not prolonged into a beak with the mouthparts at the end. 21

21. Mouthparts adapted for chewing 22
Mouthparts adapted for sucking; cerci absent; wings with few cross veins 29

22. Tarsi 5-segmented; if they are 3- or 4-segmented the hind wings are smaller than the fore wings and are held flat over the abdomen in repose; cerci absent.. 23
Tarsi 2-, 3- or 4-segmented; veins and cross veins in wings not numerous... 26

23. Prothorax small; if long then the fore legs are modified for capturing prey 24
Prothorax long, longer than head; fore legs not adapted for capturing prey; cross veins numerous .... NEUROPTERA
(part)

24. Fore and hind wings more or less similar to one another with many veins and cross veins; if venation is reduced the wings are covered with a powdery exudation.. 25
Hind wings much smaller than fore wings; wings with relatively few cells; costal cell without cross veins; abdomen usually narrowed at base to form a distinct 'waist'..................................... HYMENOPTERA

25. Costal cell in fore wing always with some transverse veinlets .... NEUROPTERA
(part)
Costal cell in fore wing without transverse veinlets .. MECOPTERA

26. Fore and hind wings of more or less equal size, held flat over abdomen in repose; tarsi 2-, 3- or 4-segmented................................... 27
Hind wings smaller than fore wings; wings usually held roof-wise over abdomen in repose; tarsi 2- or 3-segmented 28

27. Tarsi apparently 4-segmented; cerci usually very small; wings with transverse suture near base at which the wing can be fractured by insect; social insects.. ISOPTERA
Tarsi 3-segmented; cerci conspicuous; tarsi of fore legs swollen .... EMBIOPTERA

28. Cerci absent; tarsi 2- or 3-segmented........................ PSOCOPTERA
Cerci present; tarsi 2-segmented................................ ZORAPTERA

29. Wings without scales; not held spread out in repose; prothorax large;

   antennae with few segments; mouthparts adapted for piercing and sucking 30

   Wings covered in scales; antennae of many segments; mouthparts (if present) forming a tube which is coiled up when not in use .... LEPIDOPTERA

30. Mouthparts arising from back of head...................... HEMIPTERA
                                       (suborder HOMOPTERA)

   Mouthparts arising from front of head .... HEMIPTERA
                                      (suborder HETEROPTERA)

31. Mouth non-functional; abdomen with long filaments at the hind end... 32

   Mouthparts in form of a proboscis, sometimes reduced; abdomen without such filaments; hind wings represented by small knobbed processes.. DIPTERA

32. Antennae inconspicuous; wings with many cross veins .... EPHEMEROPTERA
                                                   (part)

   Antennae obvious; wings with venation reduced to a forked vein, cross veins absent... HEMIPTERA
                                     (suborder HOMOPTERA)
                                                   (part)

33. Body at least somewhat insect-like with apparent head, thorax, abdomen and legs; able to move about.. 34

   Body without obvious divisions, or without legs, or immobile...... 78

34. Terrestrial, breathing by means of spiracles; sometimes without respiratory organs 35

   Aquatic, usually with gills (larvae) 64

   Parasites on birds or mammals.. 72

35. Underside of abdomen with styles or appendages; mouthparts retracted within the head capsule; antennae present or absent; if present, the maxillary palps are less than 3-segmented..................................... 36

   Mouthparts not retracted within head capsule; if mouthparts have mandibles and are thus adapted for chewing, then the maxillary palps are more than 2-segmented; antennae always present; abdomen usually without styles 38

36. No antennae; no long cerci; no springing apparatus and ventral sucker on abdomen; tiny insects living in soil............................ PROTURA

   Antennae conspicuous; long cerci present, if not, then a ventral sucker is present on abdomen.. 37

37. Abdomen of 6 or fewer segments; a ventral sucker present near the base of the abdomen; springing apparatus usually present, small species.. COLLEMBOLA

   Abdomen of more than 8 segments; long cerci present or these replaced by pincers; blind species... DIPLURA

38. Mouthparts with mandibles adapted for chewing.......................... 39

   Mouthparts adapted for sucking 60

39. Body usually with scaly covering; abdomen with 3 long filaments at end of abdomen; never winged.. THYSANURA

   Body not scaly; never with 3 filaments at the end of abdomen 40

40. Abdomen without legs 41

   Abdomen with prolegs, differing from the legs of the thorax; (caterpillars and caterpillar-like larvae) 58

41. Antennae long and conspicuous 42

   Antennae short; larvae.. 55

42. Abdomen ending in forceps-like organ................... DERMAPTERA
    Abdomen not ending in such an organ 43
43. Abdomen strongly constricted at base into a 'waist'
    .... HYMENOPTERA
    Abdomen not strongly constricted at base; broadly attached to
    thorax.. 44
44. Head prolonged into a beak with mouthparts adapted for chewing at the
    end.. MECOPTERA
    Head not prolonged into a beak.. 45
45. Body soft; tarsi 2- or 3-segmented; usually small species, less than 6 mm
    in length 46
    Usually larger species; tarsi usually more than 3-segmented or body
    hard and cerci absent.. 47
46. Cerci absent.. PSOCOPTERA
    Unsegmented cerci present... ZORAPTERA
47. Hind legs with enlarged femora, species capable of jumping; wing buds
    of nymphs, if present, inverted, i.e. those of the metathorax lying above
    those of the mesothorax... ORTHOPTERA
    Hind legs not adapted for jumping; wing buds, if present, arranged
    normally 48
48. Prothorax much larger than mesothorax; front legs adapted for
    grasping prey.. MANTODEA
    Prothorax not much longer than mesothorax, not conspicuously
    lengthened; legs normal 49
49. Antennae usually more than 15-segmented; cerci present 50
    Antennae usually 11-segmented; cerci absent; body usually hard and
    fore wings usually hardened into horny elytra .... COLEOPTERA
50. Cerci more than 3-segmented 51
    Cerci up to 3-segmented 53
51. Body flattened, oval; head retracted at least somewhat below overhang-
    ing prothorax .... BLATTODEA
    Body elongate; head well visible from above, not overhung by
    prothorax 52
52. Cerci long; ovipositor conspicuous, tarsi 5-segmented
    .... GRYLLOBLATTODEA
    Cerci short, ovipositor absent; tarsi 4-segmented .... ISOPTERA
53. Tarsi 5-segmented, body usually very long and narrow,
    stick-like .... PHASMIDA
    Tarsi 2- or 3-segmented, body not as above................................. 54
54. Tarsi of fore legs enlarged; insects living in silk tunnels (usually on bark
    or under stones) .... EMBIOPTERA
    Tarsi of fore legs not swollen .... ISOPTERA
55. Body cylindrical, caterpillar-like.. 56
    Body at least somewhat depressed, not caterpillar-like................. 57
56. Head with six ocelli on each side.......................... LEPIDOPTERA
    (larvae)
    Head with more than six ocelli on each side............. MECOPTERA
    (larvae)
57. Mandible and maxilla of each side united to form a sucking
    tube .... NEUROPTERA
    (larvae)
    Mandibles on each side separate from maxilla....................................
    Larvae of COLEOPTERA, a few NEUROPTERA, STREPSIPTERA
    or DIPTERA

114

58. Five or fewer pairs of abdominal legs, *not* occurring on first two or seventh segment these legs nearly always ending in rows of crochets (hooks) .... LEPIDOPTERA

Six to ten pairs of abdominal legs, one pair of which occurs on the second segment; these legs not ending in crochets 59

59. Head with one ocellus on each side .... HYMENOPTERA
(larvae)

Head with several ocelli on each side .... MECOPTERA

60. Body with a few hairs or none, or with waxy coating 61

Body covered with hairs or scales; proboscis; if present, coiled up under head.. LEPIDOPTERA

61. Apical segment of tarsi swollen, small species.... THYSANOPTERA

Apical segment of tarsi not swollen, the segment ending in a claw 62

62. Prothorax distinct, usually visible from above 63

Prothorax small, usually not visible from above.............. DIPTERA

63. Proboscis arising from front of head......................... HEMIPTERA
(suborder HETEROPTERA)

Proboscis arising from hind part of head.................. HEMIPTERA
(suborder HOMOPTERA)

64. Mouthparts adapted for chewing 65

Mouthparts adapted for sucking .... HEMIPTERA
(nymphs)

65. Mandible and maxilla of each side united to form a sucking and piercing tube .... NEUROPTERA
(larvae)

Mandibles free... 66

66. Body not bearing a case made of sand, leaves, twigs, etc. 67

Body bearing a case made of various materials, into which the insect retreats on being disturbed .... TRICHOPTERA
(larvae)

67. Abdomen with lateral gills.. 68

Abdomen without lateral gills 69

Note: a few Coleoptera and Trichoptera have gills in the larval stage.

68. Abdomen ending in 2 or 3 long filaments .........EPHEMEROPTERA
(nymphs)

Abdomen ending in short processes .... NEUROPTERA
(larvae)

69. Labium modified into a large powerful grasping instrument which is folded back below head when not in use........................ ODONATA

Labium not so modified... 70

70. Abdomen without legs... 71

Abdominal legs present... LEPIDOPTERA
(larvae)

71. Segments of thorax loosely united with one another; antennae and caudal filaments long and fine .... PLECOPTERA
(nymphs)

Segments of thorax not constricted; antennae and caudal filaments short .... COLEOPTERA
(larvae)

Note: some Diptera and Trichoptera would run out here.

72. Body flattened... 73

Body compressed; mouthparts adapted for sucking, jumping species.. SIPHONAPTERA

73. Mouthparts adapted for chewing 74

Mouthparts adapted for sucking 76

74. Cerci long... 75

Cerci absent (parasites of birds and mammals)..... MALLOPHAGA

75. Cerci straight; blind; antennae short; parasites of rodents .... DERMAPTERA
(Hemimerina)

Cerci bent; eyes present; antennae almost as long as body; parasites of bats... DERMAPTERA
(Arixeniina)

76. Antennae short, but visible.. 77

Antennae not visible from above being held within pits.. DIPTERA

77. Proboscis not segmented; tarsi formed into a strong hook .... SIPHUNCULATA

Proboscis segmented; tarsi not formed into a strong hook .... HEMIPTERA

78. Legless forms (grubs, maggots or borers) locomotion by squirming or contractions of the body. (This couplet leads to many forms: larvae of Strepsiptera, Siphonaptera, Coleoptera, Diptera, Lepidoptera or Hymenoptera. If aquatic, reference should be made to mosquito larvae or pupae).. Larvae of
STREPSIPTERA, COLEOPTERA, DIPTERA, HYMENOPTERA, LEPIDOPTERA. If aquatic, MOSQUITO larvae or pupae.

Sedentary forms, not capable of locomotion................................ 79

79. Beak long and slender; often covered in wax or filamentous tufts or appearing as a small dome-shaped object attached to plants... HEMIPTERA
(suborder HOMOPTERA)

Body usually still, but able to bend; non-feeding; sometimes enclosed in cocoon or hard shell .... 80 (pupae)

80. Skin encasing legs, wings etc., prothorax small proboscis discernible 81

Legs, wings, etc. free of body; mouthparts adapted for chewing, discernible 82

81. Proboscis usually long; four wing cases present; sometimes in a cocoon.. LEPIDOPTERA
(pupae)

Proboscis short; two wing cases present; pupa often enclosed in hard shell representing final larval skin.................................... DIPTERA
(pupae)

82. Prothorax small, fused with mesothorax; sometimes in cocoon... HYMENOPTERA
(pupae)

Prothorax not fused with mesothorax 83

83. Wing cases with few or no veins.............................. COLEOPTERA
(pupae)

Wing cases with numerous branched veins.............. NEUROPTERA
(pupae)

APPENDIX

Fixing Fluids

Oudeman's Fluid

70% alcohol	88 ml
Glycerine	4 ml
Glacial acetic acid	8 ml

Lacto-alcohol

Lactic acid	40 ml
95% alcohol	37 ml
Water	23 ml

Carnoy's Fluid

95% alcohol	75 ml
Chloroform	30 ml
Glacial acetic acid	10 ml

Kahle's Fluid

95% alcohol	100 ml
Glacial acetic acid	7 ml
Formalin	40 ml

Pampl's Fluid

Glacial Acetic Acid	4 ml
Water	30 ml
40% formaldehyde solution	6 ml
95% alcohol	15 ml

KAA

Kerosene	10 ml
95% alcohol	100 ml
Glacial acetic acid	20 ml
(For very soft bodies reduce kerosene to 5 ml)	

Relaxing Fluid

Barber's Fluid

95% alcohol	50 ml
Water	50 ml
Ethyl acetate	20 ml
Benzol	10 ml

Sugaring mixture

500 g treacle
1 kg brown sugar
300 ml beer
a little rum

Boil until of uniform thickness. The more the mixture is heated the thicker it becomes. Do not boil for too long. The proportions of ingredients can be varied with experience.

Pest and mould deterrents

1. Dissolve as much naphthalene as possible in chloroform and add an equal quantity of creosote.
2. Molten naphthalene can be poured into the box or drawer and allowed to harden. *Melt in water bath*, not over open flame. Creosote can be painted on surface of box or drawer.

FURTHER READING

Textbooks

A. D. Imms, *A general textbook of Entomology*, Methuen, London, 1957. (The 9th edition revised by D. W. Richards and R. G. Davies is in some ways preferable to the later two volume edition.)

I. M. Mackerras (ed.), *The Insects of Australia*, Melbourne University Press (Sponsored by CSIRO), 1970. Reprinted 1973, with a supplement in 1974. (This is the best standard textbook on Australian insects.)

Works on techniques

A. Peterson, *Entomological Techniques*, published by the author, Ohio State University, 1959. (Difficult to obtain, mainly descriptions and diagrams of equipment for a wide range of purposes.)

K. R. Norris, *The collection and preservation of insects*, Australian Entomological Society Handbook No. 1, 1966. (Introductory booklet on collecting and collections.)

Collecting and preserving insects and their allies, Australian Museum Leaflet No. 14, 1971.

H. Oldroyd, *Collecting, preserving and studying insects*, Hutchinson, London, 1958. (An introductory book on collecting and collections.)

M. S. Moulds, *Embedding insects and other specimens in clear plastic*, Australian Entomological Press, Greenwich, 1975. (Plastic embedding is used mainly for display or demonstration purposes.)

Works on Australian fauna

I. F. B. Common and D. F. Waterhouse, *Butterflies of Australia*, Angus and Robertson, Sydney, 1972. (A comprehensive work.)

I. F. B. Common, *Australian Butterflies*, Jacaranda Press, Brisbane, 1964. (Pocket identification manual.)

R. H. Fisher, *Butterflies of South Australia*, South Australian Government Printer, Adelaide, 1978. (Deals with local fauna but has application beyond South Australia.)

I. F. B. Common, *Australian Moths*, Jacaranda Press, Brisbane, 1963. (Pocket identification manual.)

F. C. Fraser, *A handbook of the dragonflies of Australia*, Royal Zoological Society of New South Wales, Sydney, 1960.

F. Greenslade, *A guide to the ants of South Australia*, South Australian Museum, Adelaide, 1979. (Covers a limited geographical area.)

E. Matthews, *The Beetles of South Australia*, Part One, South Australian Government Printer, Adelaide, 1980.

General reading

D. Clyne, *The garden jungle*, Collins, Sydney, 1979.

A. Healy and C. N. Smithers, *Australian insects in colour*, A. H. & A. W. Reed, Sydney, 1971.

R. D. Hughes, *Living insects*, Collins, Sydney, 1974.

A. D. Imms, *Insect natural history*, Collins, London, 1947. (Revised edition, paperback, 1973.)

E. Matthews, *Insect Ecology*, University of Queensland Press, Brisbane, 1976.

Key to families of insects and related groups

C. T. Brues, A. L. Melander and F. M. Carpenter, *Classification of insects*, Museum of Comparative Zoology, Harvard, 1954.

INDEX